The Sociology of Scientific Work

T0260522

PRIME SERIES ON RESEARCH AND INNOVATION POLICY IN EUROPE

Series Editor: Philippe Larédo, *ENPC (France) and University of Manchester, UK*

The last decade has seen dramatic transformations in the configuration of national systems of innovation in Europe and in the way in which knowledge is produced. This important new series will provide a forum for the publication of high quality work analysing these changes and proposing new frameworks for the future.

In particular it will address the changing dynamics of knowledge production within the NBIC (Nano, Bio, Information, Cognitive) sciences and within different industries and services. It will also examine the changing relationship between science and society and the growing importance of both regional and European public authorities.

The series will include some of the best empirical and theoretical work in the field with contributions from leading and emerging scholars.

Titles in the series include:

Universities and Strategic Knowledge Creation
Specialization and Performance in Europe
Edited by Andrea Bonaccorsi and Cinzia Daraio

The Handbook of Technology Foresight
Concepts and Practice
Edited by Luke Georghiou, Jennifer Cassingena Harper, Michael Keenan, Ian Miles and Rafael Popper

The Sociology of Scientific Work
The Fundamental Relationship between Science and Society
Dominique Vinck

The Sociology of Scientific Work

The Fundamental Relationship between Science and Society

Dominique Vinck

Professor of Sociology, University of Grenoble, France

PRIME SERIES ON RESEARCH AND INNOVATION POLICY
IN EUROPE

Edward Elgar
Cheltenham, UK • Northampton, MA, USA

Published by
Edward Elgar Publishing Limited
The Lypiatts
15 Lansdown Road
Cheltenham
Glos GL50 2JA
UK

Edward Elgar Publishing, Inc.
William Pratt House
9 Dewey Court
Northampton
Massachusetts 01060
USA

A catalogue record for this book is available from the British Library

Library of Congress Control Number: 2009938426

Mixed Sources
Product group from well-managed
forests and other controlled sources
www.fsc.org Cert no. SA-COC-1565
© 1996 Forest Stewardship Council

ISBN 978 1 84844 964 0 (cased)

Printed and bound by MPG Books Group, UK

Contents

Figures and tables

Figures

Tables

Acknowledgements

Our thanks go to all those who helped this project to mature and to those who reviewed the various drafts. This especially concerns Michel Grossetti, Jean-Luc Guffond, Matthieu Hubert, Pablo Kreimer, Patrick Le Galès, Séverine Louvel, Marco Oberti and Ana Spivak.

Introduction

The issues facing society today (sustainable development, health and industrial risks, new technologies, the knowledge society and so on) concern science and technology. Mad cow disease, the controversy over genetically modified organisms (GMOs), nanotechnologies, our understanding of climate change, the depletion of our natural resources, the fight against new epidemics (AIDS, bird flu and so on), and the transformation of our production systems are just some of the topics that concern the human and social sciences as well as the natural, health and engineering sciences. Researchers and lecturers in these fields are making sure that students receive thorough training in these sciences (covering the state of knowledge, methods, epistemology and so on), but also on the interrelations between 'science and society'. Indeed, these are an essential key to the dynamics of science.

In science and engineering faculties just about everywhere, social science training courses have been introduced. Sometimes, the temptation is to believe that a dash of epistemology will be enough to get across to young scientists exactly what science in action is all about. Others believe that a dose of ethics is what they need to be able to deal with society-related problems. Of course, such beliefs are by and large illusory. Obviously, some kind of philosophical training has its worth, but what our young experts also need is scientific training that will allow them to get to grips with the real socioscientific dynamics. They need to be able to understand the dynamics behind the creation of knowledge and innovation, but they also need to be able to act on these, both as professional actors and as responsible citizens.

This book provides analysis frameworks to help students and scholars to decode the stakes underlying and surrounding science and technology. It looks at different ways in which science and society interrelate (for example, the emergence of scientific disciplines, the dynamics behind innovation, technical democracy and so on), and at the main social mechanisms that drive and sustain science (institutions, organisations, exchanges between researchers, building of content, concrete practices and so on). With this manual, sociology lecturers will be able to meet the rising demands of our colleagues working in the natural and engineering sciences. Its use is also recommended for new training courses such as a Masters degree in science and technology. But it will also help to prepare future generations of sociologists to deal with the science and society questions that many have tended to leave to one side.

The objective of this manual is to provide a broad range of analysis grids, concepts, methods and various other pointers about authors, schools of thought and the underlying debates. Readers of the manual will be able to understand and use Robert Merton's contribution to the institution of science and Bruno Latour's with respect to the construction of sociotechnical networks. Decoding the workings of scientific job markets sheds just as much light on science as examining the material-related and cognitive culture of a laboratory. Similarly, studying the role of language interactions in science in the making, or in scientific publishing practices or in the interactions between scientists and non-specialists, are all starting points for analysis that go beyond the contributions of epistemology or ethics. This manual does not aim to be erudite. Nor does it aim to set up or defend *one* overriding theory of science, based on rationalistic epistemology, relativism, constructivism, relationism, neo-institutionalism or whatever. On the contrary, it studies and documents the various processes and mechanisms at work, as these are highly useful when trying to understand the dynamics in action.

It is a question of understanding what 'doing science' really means. Simply detailing the state of knowledge, as is usually the case in teaching and TV programmes popularising science, is not enough to understand how such knowledge was created. A student's view of science based on what they learn from their lessons very often has little in common with science as it is practised. Even practical exercises rarely allow students to get a real grasp on research approaches. Students aiming to go into research discover the real face of science as they go along, as well as what they need to know to become a good researcher: methods, negotiating with colleagues, empirical know-how, science institutions and networks, writing styles and so on. History, philosophy, sociology, economics and linguistics all propose their own analyses. This manual has thus been written for future researchers too.

Any philosophical discourse that conveys *one* general and universal conception of science, as if it were the norm to be followed by all researchers, is counterproductive. On the one hand, it shrouds science in a mystery that is far from being compatible with actual scientific practices. Such discourse is therefore not very useful when it comes to providing researchers with concrete guidance in their work. Although it may stimulate thinking about science, and change the course of science, it is above all the privilege of those who have already proven their worth and can afford to wax philosophical about it. On the other hand, this general conception of science, which is pushed to the front when combating pseudo-science and irrationalism, is so far removed from concrete scientific practices that it loses all credibility. Without a philosophical representation that comes close to what can actually be observed or practised, reflexive researchers or outside observers are likely to fall into the worst type of relativism: 'if there is no universal science then it's all very much of a muchness'. The sociology of science, on the contrary, puts forward realistic analyses of scientific activity.

While some lecturers are afraid that the sociology of science is going to scare away their students because of the less edifying image of science that it portrays, others recommend that young researchers study it. Owing to its realistic approach, these students will become better researchers able to understand and act within the scientific world. This manual may lead some students to drop the idealistic views that had led them to pursue a career in science, while it will spark others' passion for research and the way it works. It will help the latter to adopt a more lucid approach: science and technology pose problems of an ethical, political, economic and social nature. Neither the mystifying myth of rationalism nor radical and sceptical relativism are likely to help solve these.

As well as providing scientific and sociological training, this book is for anybody interested in the knowledge society: the growing scientific controversy and the issue of expertise are prime examples of this public concern. The book outlines a series of approaches designed to shed light on the relationship between science and society.

The Turns Taken by the Sociology of Science

This manual describes different ways of studying science, but it is neither a history of ideas nor a sociological work on the sociology of science. The relationship between, for example, sociological analyses and the social engagement of their authors will only be touched on in passing.[1] Taking the sociology of science as a subject of sociological study[2] will be for another project. This kind of analysis, based on the health economy in Great Britain for example (Ashmore et al., 1989), shows how interesting it is to report on the building of research programmes, the involvement of researchers in the media, the development of instruments designed for action and the insertion of young people in society's institutions.

Approaches in the sociology of science have become increasingly diversified. The development of this field is based on ongoing *dialogue* with other social science disciplines. Philosophers have pondered over the nature of this great development over the last centuries. They have attempted to explain it by examining scientific reasoning and the intrinsic normativity of science. Historians have traced the evolution of ideas and instruments. Economists have explored the links between science and economic dynamics. The analyses performed by these various disciplines compete and contrast with each other. There are also academic quarrels within disciplines: in the philosophy of science (rationalism versus realism), psychology (different cognitive theories), economics (the neoclassical versus the evolutionist approach) and history (the inside history of ideas versus the social history of science). Furthermore, several developments in the sociology of science can only be understood by referring to the philosophy of science or to exchanges with the economics of innovation.

Nor is there any consensus as to the best way of going about the sociology of science. The diversity of approaches helps to enliven and enrich scientific production in the field. Several authors have published articles or works on its so-called 'turns': 'social turn', 'cognitive turn' (Fuller et al., 1989), 'semiotic turn' (Lenoir, 1994), 'the turn to technology' (Woolgar, 1991), 'the practice turn' (Schatzki et al., 2000) or 'One more turn after the social turn' (Latour, 1996) or Pinch's pointed criticism (1993) at the thinker Woolgar: 'Turn, turn, and turn again: The Woolgar formula'. One might also talk about the 'normative turn', in reference to the growing number of committees focusing on ethics and fighting scientific fraud.

Using the idea of a turning point is often rhetorical. The aim is either to speak out about an approach that has gone adrift or back in time (rationalist or cognitivist theories, sociological reductionism or the impasse of reflexivity), or to convince people that a major change has occurred (semiotic turn, pragmatic turn and so on). Different periods have seen different movements emerging. However, the main schools behind the structuring of the field are still active. They refer to the following representations of science:

- Science as a **social institution producing rational knowledge**: science is different from the rest of society. Its actors are scientists, critical producers of true statements, whose behaviour is governed by norms and the goal of their institution: ever-progressing knowledge.

- Science as an **exchange system**: scientific activity is geared towards nature for some and society for others. The actors are rivals, driven by the promise of rewards, by the build-up of credit or credibility or by the position that they can attain. They become rational thanks to the exchange system and the fierceness of competition.

- Science as a **reflection of local cultures and societies**: scientific activity and output are explained by social factors. Scientific activity is guided by the interests of scientists and the social groups to which they belong. The goals of science are imposed from outside. The stability of knowledge comes from the production of local social consensus.

- Science as a set of **contingent sociotechnical practices**: scientific work is linked to multiple elements (incorporated tacit knowledge, instruments, materials and so on) and results in various types of output and notably publications. The actors work in laboratories and keep up relations with society. Scientific dynamics depend on circumstances and local cognitive and material culture.

- Science as a **construction of distributed research collectives and sociotechnical networks**: scientific work consists in linking heterogeneous elements in order to produce robust entities (instruments, statements and so on). Alignment and reconfiguration mechanisms are central; they lead to relatively dense and wide-reaching actor-networks where the classic distinctions between nature and society do not apply.

The sociology of science generally switches from a study where the social aspect is seen as the central concept around which explanations are organised to other approaches where social causality is overridden by the focus on the material nature of things. The notion of science, viewed as a distinct entity, is rethought as a heterogeneous and distributed whole. Thus, the sociology of science has evolved from a sociology of scientists to a sociology of scientific knowledge, to social studies in science and technology and to an anthropology of science, technology and society.

These different analytical stances lead us to some relatively local approaches to scientific activity. Nevertheless, any globalising thinking about the relationship between science and society is rare. The sociology of science rarely raises this kind of question at the macroscopic level, even if there are calls for sociology to shed itself of some of its positivism (dissection of scientific work) in order to put forward new landscapes re-injecting new meaning into all this activity and making it possible to assess it.

Notes

1 Merton's defence of the autonomy of science in a period when the world was full of totalitarian regimes or the relativist sociologists' fight against the hegemony of physics.
2 Little work has been devoted to the sociology of the social sciences, with the exception of Deutsch et al. (1986) and Halliday and Janowitz (1992), who show that the divides between specialities are much deeper than the barriers between disciplines. There have been few efforts to draw up any theoretical summaries.

Recommended Reading

Jasanoff, S., Markle, G., Peterson, J. and Pinch, T. (eds) (1995), *Handbook of Science and Technology Studies*, London: Sage.

References

Ashmore, M., Mulkay, M. and Pinch, T. (1989), *Health and Efficiency: A Sociology of Health Economics*, Milton Keynes: Open University Press.
Deutsch, K., Markovits, A. and Platt, J. (1986), *Advances in the Social Sciences 1900–1980*, Lanham, MD: University Press of America.
Fuller, S., De Mey, M., Shinn, T. and Woolgar, S. (eds) (1989), 'The cognitive turn. Sociological and psychological perspectives on science', in *Sociology of the Sciences Yearbook*, Vol. 13, Dordrecht: Kluwer.
Halliday, T. and Janowitz, M. (1992), *Sociology and its Publics: The Forms and Fates of Disciplinary Organization*, Chicago, IL: University of Chicago Press.
Latour, B. (1996), *Aramis, or the Love of Technology*, Cambridge, MA: Harvard University Press.
Lenoir, T. (1994), 'Was the last turn in the right turn? The semiotic turn and A.J. Greimas', *Configurations*, **1**, 119–36.
Pinch, T. (1993), 'Turn, turn, and turn again: the Woolgar formula', *Science, Technology, and Human Values*, **18**, 511–22.
Schatzki, T., Knorr-Cetina, K. and von Savigny, E. (eds) (2000), *The Practice Turn in Contemporary Theory*, London: Routledge & Kegan Paul.
Woolgar, S. (1991), 'The turn to technology in social studies of science', *Science, Technology, and Human Values*, **16**, 20–50.

1 Science and society: a complex relationship

Science would appear to stand out from other social activities. This phenomenon has kept thinkers pondering, notably those striving to understand society and its transformations. Indeed, heads were being scratched well before the sociology of knowledge and the sociology of the sciences actually came into being. In this first chapter, we shall overview the analytical work of several classical authors (Comte, Condorcet, Marx and so on), who studied the relationship between science and society and, in particular, the conditions behind the presence and development of science in society. We shall study the analysis put forward by one of the first sociologists of science, Merton, who explored the relationship between Puritanism and the role of the scientist. Then, referring to the work of Ben-David, we shall look at the process according to which science emerged as a distinct social activity. Finally, we shall concentrate on the mechanisms behind the organisation and governance of the sciences in society. The question of the relationship between science and society shall be looked at again in Chapter 4 when we study the production of scientific knowledge.

Emergence of a Distinct Social Activity

In this first part, we shall see how science emerges as a social phenomenon, how the social role of the scientist is institutionalised according to the values of society, how the scientific community becomes independent of society, how the laboratory emerges as an institution and disciplines are established within society.

Science as a Social Phenomenon

The idea of science is often associated with that of a world apart, different from society. Our perception of science is still pervaded by an image of the isolated scientist, excitedly working on things beyond comprehension, or that of the genius, incarnated by Albert Einstein. Science comes across as a mysterious activity and scientists as strange beings. There seems to be a rift between the sciences and other forms of knowledge.

   Indeed, for a long time, thinkers like *Condorcet* (1743–94) suggested that the emergence of science was a specific social and historical phenomenon, with the knowledge system being dependent on the structure of society.

   For *Auguste Comte* (1798–1857), the human mind and every branch of

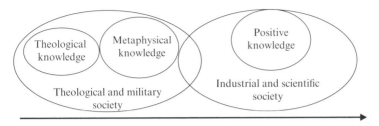

Figure 1.1 *Evolution of the nature of knowledge and the type of society*

knowledge pass through three states: theological, metaphysical and positive (Figure 1.1). In the theological state, natural phenomena can be explained by forces or beings similar to humans: gods, spirits, ancestors, demons and so on. In the metaphysical state, they are explained by great causes and abstract entities such as Nature. However, in the positive science state, the human being observes phenomena and sets up laws to establish links between them, hence abandoning the search for absolute causes. Scientific disciplines such as mathematics, physics and chemistry were the first to enter this positive state because the phenomena with which they are concerned are easier to think about. Scientific disciplines that are interested in more complex objects such as social phenomena entered the realm of positive thinking at a later, although ineluctable, date. In the positive state, scientists are able to impose their verdicts on the ignorant. These states correspond to the evolutionary stages of societies: theological and military society for the first two, and industrial and scientific society for the third. Science is thus a social and historical phenomenon linked to a specific form of social organisation in which labour is organised, in factories, in order to maximise yield and not according to custom. Moving into this stage of society supposes a dual revolution, one that is both social and intellectual and which represents a radical break in tradition.

Karl Marx (1818–83) also established a link between a social system state and a knowledge system state. For Marx, science is a historically dated phenomenon, linked to the capitalist production mode.

The Scientific Role as a Byproduct of Social Values

In the 1920–30 period, the sociologist Robert K. Merton (1910–2003) queried the cultural and historical origins of the scientific community. He described science as a sphere of social and cognitive activity that is different from other forms of activity and belief. He characterised the social climate that fostered its emergence, as well as the technical conditions that made it necessary. According to Merton, science is an autonomous sphere of activity, able to resist external influences; it defends and champions the principles of independence, discipline and pure rationality.

Merton founded his analysis on the study of the origins of the scientific community in seventeenth-century England. He analysed the biographies of the members of the British elite, the activity of the Royal Society (founded in 1645) and various works, inventions and publications. He underlined the significant growth in technical knowledge, skills and machinery in the mining, metallurgy, shipbuilding and weapons industries from 1620 onwards. Merton specifically focused on the values, beliefs and feelings that marked this period of rapid development in science and technology.

Performing a quantitative study of the changes in career choices of the English social elite, he observed an evolution: in the first half of the seventeenth century, the 'science' and 'medicine and surgery' categories became increasingly popular. The elite turned more readily towards science than to the army or navy, or to the arts (painting, sculpture, music, poetry and prose), education, historiography, religion, scholastic knowledge, law or politics. According to Merton, this phenomenon could be explained by the enhanced value of the social role of the scientist and by a form of social recognition of science as an activity. At the time, there was a convergence of values between those of English Puritanism (interest in earthly things, discipline, condemning of idleness, free examination and distancing with respect to traditions and utilitarianism), and those arising from naturalist philosophy and experimental science.[1] These values, which placed experience at the top of the hierarchy of forms of knowledge, influenced the founders of the Royal Society. They permeated the Baconian movement (as of 1640) and were embedded in scientific education. The convictions at that time, with respect to man's mission to the relief of man's estate, converged towards the idea of a better understanding and control of nature. The idea of a natural science, studying the order and regularities of nature, was associated with the virtues of a new profession dedicated to it. For Merton, the growth of science as a distinct sphere of activity and the emergence of a new professional role in society could be explained less by the incoming flow of new knowledge than by the trend in social values and by the attempts of the members of the Royal Society to justify the ways of science before God. The Puritan values, combining rationalism and empiricism, fostered scientific method and gave a new lease of life to the empirical science that had been decried in the Middle Ages.

Merton's conclusions, which were similar to those of Max Weber regarding the growth of capitalism in Germany, led to the idea that the development of science is conditioned by an emphasis on the religious value of certain activities. This did not immediately lead to the institutionalisation of science, still considered an esoteric activity that was potentially dangerous for those in power and whose practical use had not yet been proven. The emphasis on religious values created favourable conditions for the development of science and the new social role of the scientist. This theory is opposed to the common idea according to which the recognition of science in society comes from its ability to solve problems. This is not at all the case; the appearance of modern science can be explained by the social values that psychologically restrict individuals. The social role of the scientist is defined by a set of behavioural norms.

The Scientific Community as the Fruit of Autonomisation

The social role of the scientist emerged simultaneously in France and Italy. The sociologist Joseph Ben-David (1920–86), in *The Scientist's Role in Society* (1971), suggests studying the history of universities in order to understand the phenomenon and the speed at which this new social role spread outside of England. In fact, scientific training was already organised in Universities independently of the powers of Royalty and the Church. Sometimes academics formed corporate bodies with their own working rules. In Paris, the university had several hundred lecturers. As secular scholars, working inside a medieval university, their role was to search for truth and criticise the ideas of their peers, while their behaviour was controlled by their intellectual community. The scholar's social role came into being without being attached to any form of power. Scholars were able to compare and contrast their ideas because they were not absorbed by the need to constantly justify their role in society. With the university, erudition had become a rightful vocation and occupation. In this social space, dedicated to education, the practice of debating and querying fostered the right conditions for autonomous research. In this way, philosophers were able to gain independence from religious authorities. The new scientists used this autonomous university structure to reproduce the same kind of system while being careful to underline their difference with respect to the philosophers. They set up informal meetings and private lessons (notably at the Collège Royal de France), and this eventually led to the establishment of scientific academies.

In Italy, scholars formed alliances with artists and engineers and strove to solve problems by combining knowledge of classical texts, practical experience and explanations of the principles at work in various phenomena: perspective in architecture, dynamics in machinery, anatomy and so on. They served as an intellectual and social resource for artists and engineers. They were also admitted to the princely courts. Between the fifteenth and seventeenth centuries, groups of scholars travelled across Europe looking for settings to match their ideal of society. They saw experimental philosophy as a means of increasing their knowledge of man and nature. The coming together of the interests of these groups and those of their hosts can explain the emergence and recognition of their social role (Box 1.1). This role consisted in studying nature using mathematics, measurement and experimentation, rather than interpreting texts in order to study divine or human ways. This is how the science academies came into being at the beginning of the seventeenth century, in particular the Academia Dei Lincei (1603) and the Academia del Cimento (1651).

It has to be said that the English revolution occupies a specific place in the history of science. It led to the merging of scientism and puritan religious values and beliefs. This merging provided a legitimate basis for the recognition of science, its role in society and its value. For the first time, the role of the scientist was institutionalised as a distinct social role making ongoing research possible. From an individual and self-taught activity, experimental science was transformed into a recognised and collective activity. The creation of the Royal Society in England (1662), followed by the Académie des sciences in France (1666), were part of this movement to institutionalise science. Scientists appeared to the rest of society as a homogeneous community, governed by rules (a normative structure) and an internal social control system. They demanded that their role and autonomy in society be recognised. Their use of mathematics helped them to stand out from other intellectuals and their doctrine-based approaches, as well as dilettantes and charlatans. The scientific community was built up outside the universities, which were still highly influenced by traditional disciplines. Nevertheless, having no specific institutional reproductive mechanism, the community still had to depend on these universities.

In different European countries, the scientific community under construction also claimed to be neutral and autonomous. It carefully selected its members so as to increase this autonomy further and partially isolated itself from other institutions, notably the universities, which it criticised. The community built itself up by excluding amateurs. The academies became the arenas of public debate where scientists reviewed scientific work. Furthermore, the scientific community gained the support of various authorities, depending on the country, and set itself up as an international and autonomous community.

Box 1.1   *Several key notions put forward by Ben-David*

Role: what is expected of an individual, a group or an institution acting as a unit of all systems making up society.

Reference group: a group holding specific expectations and sanctioning the behaviour of its members. The group grants its members retribution in the form of a *status* within the group. Members' motivation to fulfil their role depends on the retributions granted to them by the group. When members are subject to expectations and retributions emanating from other systems, they are faced with a *role conflict*. An agent can nevertheless *combine roles* and, hence, make innovation possible.

The Move from the Charismatic Scientific Revolution System to the Institution of Laboratories and Disciplines

In the eighteenth century, science was practised here, there and everywhere: in the academies, in the royal courts, in a few universities in the north of Europe and in individuals' homes. It developed around charismatic leaders, but lacked organisation. It experienced bouts of renewed esteem and bursts of creative activity, according to religious and philosophical fashions, but was not organised so that young people could be trained; it lacked continuity. Given this context, the appearance of new theories in physics and chemistry, combining experimentation and mathematics, did nothing to change the social structure of science.

In the nineteenth century, science was returned to the midst of the universities for contingent, political reasons rather than scientific reasons. There was a movement to regenerate the universities. The protagonists of this movement, both philosophers and scientists, attempted to set up professional training for the benefit of the state. In France, as in Germany, they deplored the universities' backwardness and the authoritarian criticism that they practised. Being associated with oppressive institutions, the sciences in France were attacked by revolutionaries. At the start of the Revolution, the pure sciences and mathematics were considered as antisocial and aristocratic, before being reinstated as part of a philosophy based on the idea of Progress. The French *École Polytechnique* and *École normale supérieure* created within this context were part of this new type of academy in which the nation's scientific elite joined forces with the most outstanding scientists in order to dispense cutting-edge professional training in which science played a major role. Following the theories of Antoine Lavoisier and Pierre-Simon Laplace, which reflected the ideal of a perfect merging of mathematical theory and empirical data, defended by the academy, professors of science dispensed with knowledge that carried just as much weight as Latin grammar. Scientific education was recognised, but this did not automatically lead to training in research. Research remained within the academies (where scientific discussions took place) and private laboratories. Scientists helped each other in order to safeguard the academy of science. Science seemed to embody perfection, which was something that was so far lacking in biology and the social sciences. Classical physics was seen as a model and doubt was shed on any research that did not comply with this model.

In Germany, civil servants and philosophers joined forces to create the new Berlin University. It was designed as a grand academy, where creative scholars came together but where science was relegated to a minor role. This was because the philosophers thought they had already gathered together everything that deserved to be known into a philosophy of nature. In France, as in Germany, nobody seemed to be concerned with organising research.

However, the ambition to transform the Berlin University into the centre of German intellectual life (in the wake of political and military defeat due to the Napoleonic invasions) motivated other German states to reform their universities. Teaching became a political priority (the *Gymnasium* teachers had to be trained), and primacy was given to the faculty of philosophy (which included literature and science) over the other faculties (theology, law and medicine). The country's universities did not want to lag behind and offered professorships to attract the best young scholars while the slow economic development drew many gifted students to university careers. Alongside the laboratories, designed as supportive educational structures for teaching physiology and pharmacology, the seminars organised by the universities to improve their students' skills became places of research and experimental scientific practice. However, such practices were still considered to be beneath the universities. Rivalry led to laboratories being created in order to attract students rather than for research as an end in itself. Towards 1825, a network of laboratories within the universities emerged, although this was not in itself one of the objectives for science. Without necessarily having any scholars of great genius, this research organisation, composed of competing laboratories, led to some tangible results and a high level of scientific productivity, exceeding that of France. In around 1860, the scientific disciplines practised in these laboratories began to result in practical applications in chemistry and medicine, followed by electricity. The university research system was then protected by the German states and copied in Great Britain, France and the United States. Just about everywhere, research attached itself to teaching. Nevertheless, these transformations were not fostered by any kind of intellectual movement. Once they had occurred, teaching and research were considered to be naturally linked and science as being useful.

These universities formed a subsystem, although society did not expect this subsystem to fulfil any significant economic function. It was therefore relatively autonomous. With most positions teaching literature reaching saturation between 1830 and 1840, young scholars turned towards the empirical sciences and asked for chairs to be created for the new disciplines. As only a limited number of chairs were created and then occupied by professors throughout the 30 or 40 years of their career, younger candidates turned towards other specialities and thus encouraged the creation of new disciplines.

Up until then, science had taken on two institutional forms. First, there was that of the homogeneous scientific community, governed by rules, benefiting from an internal social control system that was autonomous and recognised by society. Its model was that of the academy of science as a place of scientific discussion and peer recognition. The scientist's role was associated with this form. Its success stemmed from its ability to impose a cognitive trend that excluded metaphysics. However, it was incapable of producing the science that it promoted in a meaningful manner and was likewise incapable of reproduction. It is nevertheless this institutional form of science that the first sociologists of science focused on. The second form is that of the university laboratories, which competed with one another, were associated with teaching and whose success was driven by the fact that they attracted the best students and produced practical applications. These laboratories formed a partially autonomous system and eventually became the preferred institutional form for scientists the world over. This preference is reflected in the faster development that took place in both fundamental and applied research.

Local Organisation/Transnational Evaluation Dynamics

In French education, scientific training was more like indoctrination based on established theories, codified methods and exemplary cases. Research subjects were defined by the teacher as a set of problems to be solved within the framework of an established conceptual system. In Germany, scientific achievements enabled young scholars to pick up well-paid teaching positions. They were then able to exert control over the resources devoted to research. This led to the setting up of research schools that campaigned to have their speciality recognised as a discipline. Discoveries that did not fall within recognised disciplines or herald the creation of a new discipline were ridiculed and denied university chairs. The nineteenth-century German university system was characterised by the building of frontiers. As the German competing laboratory system had been copied elsewhere, the disciplinary conception of science spread, especially in university teaching.

The way science was organised clashed with that of the academy, seen as a form of superior authority dating back to the seventeenth century. The academy appraised scientific discovery according to formal and general criteria, taking an unbiased stance with respect to the issues under study. However, with the birth of different disciplines, some very specific theories were forged. These carried with them implicit (metaphysical) facts about the nature of the object being studied. There was no superior, controlling authority beyond the disciplines themselves. Diverging discoveries and research directions developed. Because science was carried out in so many different places and there was a tremendous traffic of ideas, it became possible to override resistance from the established schools of thought or disciplines. Although the disciplines holding the monopoly delayed the granting of resources to new scientific trends, at the same time this urged scientists to seek recognition beyond the university chairs, that is, through publications, international conferences and review by the academies. While some academies gradually lost much of their influence (for example, Berlin and Paris), others came to play an important role at world level. This was the case of the London Royal Society, and the Science Academy in Sweden with the setting up of the Nobel Prize in 1901.

Thus, the world of science was swept by a combined movement for differentiation, decentralisation and the search for consensus. According to Ben-David, the organisation of research and training in research took place within the universities, schools and laboratories, generally according to discipline, while evaluation, control and the granting of recognition were the business of reviews, the academies, scientific societies and conferences. In the eighteenth and nineteenth centuries, scientific societies and academies of science sprang up and their collective opinion was accepted as an authority. Hitherto considered to be the supreme authority, the academy was replaced by a multitude of mechanisms designed to assess work and allocate resources. These mechanisms applied judgement norms similar to those of the former academies, with the difference being that more weight was given to empirical back-up than to a link with an existing theory. The forming of new scientific consensus was no longer monopolised by a specific institution but the end result of different appraisals. Because the institutional framework of the sciences transformed these into a distinct activity, driven by its own authority, it was reinforced.

Ben-David put forward a model for analysing the dynamics of science based on two factors: (i) the competition within the decentralised and multinational academic system (the most appropriate system for developing researchers' professionalism being the decentralised and competitive American system); and (ii) the transnational assessment mechanisms which, unlike the research organisation, were not seated in a local and national context. These independent assessment mechanisms offset the fact that free and non-utilitarian research was dependent on specific authority systems at a local level. This model leads to the hypothesis that scientific research is independent of society's values.

Societal Regulations

We have just seen how the scientific research activity emerged and gained auton-omy. We shall now examine the way this activity is linked to the political and eco-nomic authorities of the twentieth century: its relations of interdependence, the way it changes as society itself is transformed and the different institutional forms that come into being according to the country. The aim will be to identify some of the ways in which society regulates scientific activity, that is, the mechanisms and interactions that ensure a balance is maintained or, on the contrary, upset the equilibrium. Finally, we shall look back over the last 30 years of European research and conclude with some of the major questions that an observer may raise.

The 'Republic of Science'

By the end of the nineteenth century, science was mainly practised in universi-ties, but also in companies that had equipped themselves with industrial research laboratories. The role of the research-dedicated scientist was recognised. With evaluation mechanisms being transnational, they were relatively independent of local political regimes. Science had also become an object of national pride. It was sparked by the competition for prestige. Nations claimed to be the fathers of inventions and disciplines. Lavoisier, for example, wanted the identity of his discipline to be sealed with that of his nation: 'chemistry is a French science'. The Nobel Prize also became the object of competition between countries.

Together with these nationalistic tensions, there emerged a trend for inter-nationalism. Scientists became accustomed to meeting each other in national meetings (first scientists' congress in France in 1864) and international confer-ences (first conference on botany in 1864, on ornithology in 1884, on physiology in 1889, on chemistry in 1894 and on mathematics in 1897). International socie-ties were created (on seismology in 1903, on solar energy in 1904, on astronomy in 1909, on geography and geology in 1922 and on radiation protection in 1925).

Some nations, such as Germany, placed great hope in science and technol-ogy. Some researchers, whose works led to practical applications, were enrolled to work for military organisations or to further industrial development. Thomas Edison created Menlo Park in 1876; Höchst and Agfa recruited chemists in1875; Bayer had around 15 working for them in 1881; Kodak in 1886; Standard Oil in 1889; Du Pont de Nemours in 1890; General Electric in 1900; and Westinghouse in 1903. In 1914, France created the Commission supérieure des inventions intéressant la Défense Nationale (superior commission for inventions concerned with national defence). The United States set up similar commissions.

After the First World War, science was surrounded by a social atmosphere dripping with disillusion. Germany had to face sudden failure. This undermined the association that had been built up between science, industry and the power of the nation. The population accused scientists of being the nation's downfall; scientific institutions were threatened while there was a strong move to return to more romantic and spiritual values. Moreover, the belief in ongoing human progress based on science was shattered. Science and scientists were at times publicly decried. There was talk of a 'moratorium on inventions' (Barber and Hirsch, 1962). The West was swept by economic recession and scientists experienced a period of academic unemployment.

Relations between the sciences and scientists fluctuated between interdependence and autonomy. In spite of this crisis, nations began to reinvest in science. New organisations were created: the National Research Council in the USA and the National Advisory Research Council in the UK in 1915, followed by the UK Department of Scientific and Industrial Research in 1916. In France, the first public body for applied research, the National Institute for Agronomic Research (INRA), was set up in 1921 then, in 1922, the National Office for Scientific and Industrial Research and Inventions (ONRS) was created. These were followed by the National Centre for Scientific Research (CNRS), dedicated to fundamental research, in 1939. In the 1940s, many scientists did everything they could to convince their governments to finance research. The states gradually set up new institutions for science in which researchers could work; these were instruments of science policy that allocated research resources. As for industrial research, this became an economic stake in the 1930s; 52 per cent of American companies claimed to be involved in such research, employing 33,000 people to this end.

Nevertheless, in the 1930s, as Nazism, fascism, communism and other totalitarian regimes were springing up across the world, scientists began to ponder their relationship with society; they queried their social responsibility. Some became involved in the preparations for a new war while others (Jews and leftwing supporters, on the one hand, and communist dissidents on the other) were persecuted. Between 1933 and 1938, 1,800 German scientists were expelled from universities. International scientific mutual aid was organised[2] while a new type of discourse emerged proclaiming science to be a spiritual enterprise, an enterprise that was democratic and autonomous (theory defended by the sociologist Robert K. Merton), and removed from nationalistic interests. Science had to be protected against social pressures and hence avoid involvement in political affairs. The autonomy of science was presented as a defence of democratic principles and science was invited to set an example, as a model of democracy. This nevertheless caused much debate as scientists feared they would be locked away in an ivory tower (involved in activities far removed from the realities of society).

With the Second World War, scientists across the globe, whether liberal, leftwing or rightwing supporters, joined forces to help the allied governments improve the performance of their military machine in order to defend the freedom of the nations.[3] However, several scientists did not want to dirty their hands and created the Society for the Freedom of Science. They wanted to set up a 'Republic of Science'.

In the United States, following the Second World War, Vannevar Bush battled to establish solid foundations for a research activity that would be driven by curiosity alone. In surveys on the prestige awarded to professions in the United States, scientists occupied 8th position in 1947 and 3rd position in 1963.

Planning Science

In France, the postwar elite became aware of the backwardness of their science and technology compared with the Americans and the British. In an effort to make up for lost time, they created new applied research bodies, including the Commissariat pour l'Energie Atomique (French atomic energy agency) in 1945. One scientific entrepreneur in Grenoble, Louis Néel (Pestre, 1989), a passionate scientist and university professor fascinated by technology, with no particular political ambitions, built up an empire while his local organisational innovations were used as a model for research institutions. Others, *'planistes'*, such as Jean Perrin, advocated the drawing up of a national research and development (R&D) plan, including the definition of themes, objectives and priorities. They organised expertise, subsidised research and set up links between laboratories. Still others, promoters of logic based on independence and national pride, such as Pierre Guillaumat, attempted to organise a prestigious scientific and industrial complex. Finally, the protagonists of industrial-based logic, such as Maurice Ponte from the company TSF, focused on developing the electronics sector. The organisation of research was the subject of political debates while substantial resources were devoted to it. The notion of planning science differed from one country to the next. In the United States, it involved mobilising the scientific community around key objectives, and coordinating and managing efforts. In France, it was more concerned with the centralised management of resources.

From then on, the sciences were considered as strategic resources for industry and the independence of nations. In the United States, the share of GNP allocated to research went up from 0.3 per cent in 1940 to 3 per cent in 1965. In absolute terms, this meant that total R&D spending was multiplied by a factor of seven and federal spending alone was multiplied by a factor of 200. Requiring a substantial amount of resources, the sciences increasingly depended on support from society. Science policy became a subject of debate. In the 1930s, this debate notably opposed the Marxist crystallographer John D. Bernal (1939) and the sociologist Michael Polanyi. For Bernal, it was a question of organising and planning researchers' working environment in order to encourage creativity. Polanyi (1958), on the other hand, thought that researchers should be left alone given that fundamental research is not something that can be steered. He believed that new fields see the light of day thanks to tried and tested individual initiatives. Trust should therefore be placed in the informal mechanisms of the scientific community.

The considerable support for research nevertheless buried this debate until the 1960s and Bernal's viewpoint imposed itself. The issue at stake was not whether the state should or should not plan science, but how science should be planned. In France, science policy insisted on the need to identify original subjects and support these. It recommended research that could be transposed into tangible scientific productions, sometimes thought of as gadgets such as the laser or the maser. At the same time, Organization for Economic Cooperation and Development (OECD) and the French Atomic Energy Commission (CEA) research specialists hammered away at concepts such as 'technological lag' and the 'ratio of GNP research expenditure'. They defined a vocabulary and indicators, listed in reference texts called the *Frascati Manual* and the *Oslo Manual*, in order to compare the 'research effort' of different nations.

However, after the 1950s, the development of the research system slowed down while the enormous investments that had been agreed for technological programmes were disputed in terms of both their relevance (nuclear energy, genetic engineering) and their performance. The output of research appeared to be less than proportional to the investments made. Moreover, the sciences had grown so much that nobody could really grasp them in their entirety. The historian Derek de Solla Price (1963) pointed out that the 50,000 scientists identified at the end of the nineteenth century had multiplied to reach one million. Different reviews abounded: increasing from one hundred in 1830 to several dozen thousand. The world of science seemed to be increasingly fragmented.

During the same period, the popularity of international scientific cooperation programmes grew. Science became 'Big Science' with all its 'heavy machinery', such as particle accelerators, managed by international organisations. The European Organisation for Nuclear Research (CERN), in Geneva, employed 3,000 people, out of which 300 were physicists. Twenty different countries contributed to the organisation's budget, amounting to roughly €500 million in 2000. At European level, logic based on integration was at work. This meant the building of large-scale pieces of equipment and the setting up of big laboratories such as the CERN, the Joint Research Centres of the European Community Commission, the European Molecular Biology Laboratory (EMBL) and the European Synchrotron Research Facility (ESRF). This logic was furthered by the creation of big international programmes: European programmes, international cooperation in the Antarctica region, the 'Human Genome' project and so on. These programmes fostered the emergence of scientific cooperation networks (Vinck, 1992). The laboratories were often too small to alone solve the scientific questions raised by epidemics such as AIDS or mad cow disease, or the planet's changing climate. At the end of the twentieth century a dual movement had taken shape combining the creation of scientific and technological skills clusters and major laboratories, on the one hand, and the creation of scientific cooperation networks between both private and public laboratories, often involved in complementary specialities, on the other.

Far-reaching mutations were affecting the organisation of scientific labour. Science had become less than ever the work of isolated researchers. The number of articles signed by a single author was halved between 1920 and 1950, while there was an increasing volume of papers co-signed by at least four authors. Science was increasingly about organisation and collective, international dynamics. The notion of science policy itself changed. From the idea of planning, it moved on to questions relating to the development, renewal and evaluation of scientific potential and research infrastructure. Within this context, 'basic researchers' were no longer scientists mastering knowledge and implementing the right scientific method. They were no longer scholars claiming individual autonomy. Modern researchers worked in teams, in a laboratory and as part of a network, on projects that they learned to organise and manage.

Plurality of National University Systems

Institutional and organisational forms of research vary from one country to the next. They are also regularly called into question. Indeed, the question of whether or not to abolish universities has been the subject of discussions since the eighteenth century. This is because they are considered to be relics of a feudal, corporatist past, delivering training that is too general and inefficient in an attempt to cover everything from cutting-edge research to education of the masses in all disciplines, to patent filing. Over the last two centuries, right up to the present day, reformers have suggested replacing them with specialised schools (technological and professional university institutes in France), whose aim is to prepare students for their future profession, using more effective teaching methods and keeping preparation for a career in research for a minority. This trend is nevertheless not shared by all. New universities continue to be created worldwide. All develop teaching of professional practical know-how, sometimes to the detriment of a disciplined and conceptualised approach. They wear themselves out teaching the masses while at times neglecting research. Such critical evaluations are at the heart of recent attempts to reform European universities, with the debate oscillating between the idea of reinforcing the complementarity between teaching and research or setting up specialised functions in specific institutions.

Specialisation is another topic of debate. Ideas about the universality of knowledge and the unity of the sciences should lead towards the development of high-level general training. Some therefore suggest introducing a bigger common-core syllabus or letting students follow courses covering a very broad range of disciplines. However, the fact is that training sectors have become specialised, universities have been split up into different faculties and students, even when they can put together their own training programme, tend to specialise at an early stage.

National university systems differ greatly when it comes to these questions of balance (research work/teaching, universality/speciality). We shall take a look at the cases of France, Germany, Great Britain and the United States.

France
With the French Revolution, a clear distinction was made between two types of establishment: (i) the universities and *Grandes Écoles* such as the *École Polytechnique* and *École normale supérieure*; and (ii) the centres for intellectual work such as the Collège de France, the natural history museum and the École pratique des hautes études. Although universities were somewhat marginalised, the system was centralised in spite of the many different establishments. In 1875, the takeover of the republicans resulted in the faculties becoming relatively autonomous. The various ensuing governments gave a certain amount of leeway to local and private initiatives. Cities began to compete in an effort to create new,

locally financed chairs and in order to attain university centre status. The possibility for industrialists to finance teaching meant that some became interested in university questions. Science faculties created technical institutes, which gradually brought the differences between university cities to the fore (Grossetti, 1994); universities sprang up wherever there were alliances with local authorities and industrialists, as in Grenoble (Box 1.2), Nancy and Toulouse.

Box 1.2   *The case of Grenoble*
In 1870 the Faculty of Science was barely active. However, when the first dams and hydropower plants were set up, the town (25,000 inhabitants) underwent a spurt of growth. As of 1876, a number of scientific personalities specialising in electricity and chemistry got involved in local politics and created a 'public evening course' on industrial electricity. The course was opened at the Faculty of Science and subsidised by the Municipal Council. In 1889, the university decided to devote most of its resources to this field. A society for the development of technical teaching, linked to the university, was founded in 1900 and the Electrical Engineering Institute was opened the same year.

Aside from the university, the *École polytechnique* also trained 'Corps' engineers: high-level civil servants in charge of planning and managing major public and military works. Their training was based mainly on encyclopaedic knowledge and the fundamental sciences, that is, 'understanding why something works'. In 1829, modernist business men created the Ecole Centrale des Arts et Manufactures (central school for industrial arts and manufacturing) and a civil engineering diploma to teach students 'how to get machinery and plants to work' (Grelon, 2001). However, the engineers that graduated from this school, like those graduating from the *écoles polytechniques*, abandoned the industrial world for prestigious state-paid positions. At the end of the nineteenth century, the second wave of industrialisation generated a new demand for qualified technicians to organise companies working in the chemical, iron and steel and electrical engineering sectors. There was thus a need for new schools to be set up. The *écoles d'arts et métiers* (industrial arts and crafts colleges) began to supply companies with graduates although the same *grandes écoles* trend could be seen: strengthening of scientific content and reversal in theory versus practice ratios. The picture was finally completed with the setting up of *instituts polytechniques* by the faculties of science in order to train engineers to work with new technologies.

After the Second World War, the state took back control over the institutions and centralised their management. Reforms were decided without consulting local authorities. A single type of recruitment method was defined while programmes and diplomas were harmonised. It was time to rebuild the nation, to harmonise and to rationalise. Large bodies were created to boost research, including the CNRS, but were external to the universities and *grandes écoles*. Nevertheless, it was more or less the same intellectual and scientific circles that held the university chairs, coordinated research at the CNRS and enjoyed the highest distinctions:

appointments to the Académie des Sciences and Collège de France. The value of university appointments was undermined and the worst thing that could happen was to be appointed to a university out in the country. The conservative traditions of the university circles basically continued to predominate.

Although most research and teaching assets were to be found in Paris, at the CNRS, the CEA, in the *grandes écoles* and Parisian universities, several applied science disciplines began to emerge together with engineer training far away from Paris. They benefited indirectly from the Parisian focus on 'fundamental research' and its lack of interest in applications. Their development was also helped by the links created between university research and industry. The presence of teachers of applied mathematics linked to engineering schools transformed cities like Grenoble into the first university centres to focus on information technology (Grossetti and Mounier-Kuhn, 1995), alongside electrical engineering and automation control. The processes at work combined local dynamics (scientific, industrial and political) and actions with national influence (alumni networks, political affiliations, informal coordination between university vice-chancellors or heads of engineering schools) (Box 1.3).

Box 1.3   *The case of Grenoble (cont.): organisation of action at local and national levels.*

Various personalities, including P. Mendès France and the industrialist P.L. Merlin, deplored the failings of the French higher education system: lack of relationship between research and teaching, lack of dynamics with respect to new disciplines, lack of engineers and managers, and the system's generally elitist approach. A movement emerged, developing into a conference in Grenoble in 1957; this gave birth to the *instituts universitaires de technologie* (IUTs, or university technology institutes).

In the 1960s, a national land development policy (decentralisation) was launched. Local actors got involved. Scientists made sure that there was a link between what was happening at national level and what local authorities were doing. In 1955, Louis Néel persuaded the CEA to create a local research centre. Conversely, local inventions, such as associations between university laboratories and the CNRS or the creation of research contracts with the socioeconomic milieu, took on a national scope thanks to networks of researchers. Again, at local level, other innovations were seeing the light of day: the idea of part-time contracts at the university for people working in a company so that there could be better interfacing between the university and companies.

Germany

The French university system is radically different from the German system. Until the Second World War, the German university was seen as a model: it was the birthplace of many disciplines, science was an end in itself and the position of university professor was seen to be highly prestigious. The general rule was that each discipline be represented by only one professor within the university, which led to the number of chairs and therefore disciplines growing considerably. This movement to create and institutionalise disciplines nevertheless quickly slowed down. At the start of the twentieth century, the university system seemed to have reached a stalemate situation. The university structures put the brakes on the growing number of specialities. In less than one century, the university went from being a dynamic and creative institution to a frozen structure. University lecturers protected their institution against the risk of invasion by practical subject matters. Society and the political regime adopted an absolutist approach, expecting the university to train the country's elite civil servants, judges, prosecutors and teachers, in exchange for which the universities were free to choose their research subjects. Emerging sciences, such as social and engineering sciences, were not considered worthy of being taught at university. It was primordial to protect theoretical teaching, pure research and the idea that each discipline had its own methodology. It was in this context that the institutes were created, considered as annexes designed to facilitate research practice for professors, without undermining the organisation of university chairs. Work geared towards companies and military applications was performed in these institutes, without this necessarily leading to the creation of university chairs. The engineering sciences, excluded from university curricula, were located in separate establishments where they developed until the state directly intervened to transform them into technological universities.

Great Britain

The British university enshrined two traditions: provincial universities, created by the middle classes in towns, and the great universities of Oxford and Cambridge, where nobility and Churchmen received their education (character and lifestyle training rather than practical training). During the first shift in the higher education scene, the University College of London was the first university to be created with a view to providing practical training, notably in medicine. This was then backed up by the University of London, which bestowed recognised diplomas on students from diverse social backgrounds, provincial colleges and the colonies. Both university categories met the demands of different social milieux: the first offered a broad range of professional teaching with research being geared towards local needs; the second trained the elite, providing just a few openings for research and professional teaching.

Although the German university had been used as a model to reform Oxford and Cambridge and create research laboratories in the 1870s, the operation was not a success. Being prosperous and prestigious establishments, Oxford and Cambridge taught an elite population of students aspiring to literary, scientific or political creativeness. These students did not have any immediate professional needs although they sometimes went on to gain further professional skills in specialised schools. The training dispensed at Oxford and Cambridge above all focused on the first few years of study. Owing to a lack of space in which to extend their laboratories and because the universities were not in competition with other establishments, there was little development in terms of research. As for the provincial universities, which specialised in technical, agricultural and commercial training in line with local needs, these recruited young graduates from the more prestigious universities. Having graduated from Oxford and Cambridge, they taught what they had themselves been taught. This led to education being standardised from the top of the university hierarchy and to an absence of competition based on differentiation. The lesser universities were seen to be prestigious when they imitated Oxford and Cambridge, which were not required to be innovative.

The engineering sciences, which were allowed to be present in British universities unlike in German universities where they were excluded, remained marginal (as was the case of Imperial College). The flexibility of the British system thus led to professional, engineering and also writers' training being included in university curricula. The range of posts and statuses meant that much variety could be integrated without there necessarily being any formal recognition of a new field.

United States

American universities were born from the British tradition, but in a society where people, including those from different social classes, were much better off than in England. In the nineteenth century, the United States boasted many colleges, often religious, as well as professional training establishments, none of which was used as a model. Given that research had no proven use, only a marginal amount was carried out in these establishments, in spite of it being defended by scholarly circles and enthusiastic amateurs. Erudition and science were, at best, a pastime. As of 1860, practical applications began to appear and professional establishments (notably the Massachusetts Institute of Technology: MIT) and universities (including Johns Hopkins University), with a strong scientific base, were created while others were reformed (Harvard). Because there was no one university providing exclusive training for the elite, each one competed with the others in an effort to meet the rising social demand. The older universities were not in a position to stay out of this movement. The model created by the young Johns Hopkins University (that is, a 'research university') was followed by Harvard. The dynamics behind this competition and the urge to copy what was being done elsewhere contributed to the development of professional university training or the introduction of courses in literature and the social sciences. The big universities covered the first few years of study and provided professional training, research-based post-graduate degrees and research institutes. Owing to the competition, they sought to be different and hence offered new services: new specialities and branches of training, activities geared towards sports and the arts, alumni associations and so on. They specialised according to their assets; instead of systematically covering all disciplines, they poured their resources into what they thought had potential. The universities began to look more and more like businesses, operating on selected market segments, rather than institutions that were supposed to represent all knowledge. Also, if students wanted to pursue a PhD in a specific field, their choice of university was restricted to a small, constantly changing, list.

The American university was made up of departments where there were several professors, sometimes working as a team in order to redefine the collective strategy, and a team of PhD students; the status of teachers, some of whom were only taken on temporarily, differed from that of researchers who hardly ever taught, teachers with a certain reputation, professionals who dispensed courses in their own particular field and specialised administrators. Although flexible, this type of organisation was also fragile as it was highly dependent on the social demand that it helped to raise. Overall, the system had a high capacity in terms of meeting heterogeneous demand and thus growing outwards, but this also meant substantial differences between establishments, functional hierarchies, levels of prestige and quality of training.

National innovation systems

When thinking about the diversity of scientific systems, innovation economists put forward the notion of a 'national innovation system'. This refers to the institutional differences at work in the production of knowledge and innovation. It takes into account the training system, the way research is organised, the rules governing job markets, state intervention and industrial and financial organisation. It also covers 'social innovation systems' (sets of routines, procedures and institutions governing innovation-related behaviour). These economists identify four innovation systems (Amable et al., 1997):

- **Market system**: institutions are mainly organised around the market, which acts as a vector of adjustment. The state is fragmented into agencies. In this system, research is based on competition between institutions and individuals. Patents and copyright play a role in terms of inviting invention (individualisation of the benefits of innovation). Private innovation trends, geared towards product renewal, are only slightly modified by large public programmes (defence, space and so on). Venture capital is active. Training is for highly qualified people, who contribute to innovation, and those more involved in production (United States, Canada and Australia).

- **Integration system**: public institutions are at the centre of innovation and regulation dynamics. They fashion economic circuits (supply and demand): homogenisation of markets, adoption of common rules and development of social coverage. Competition is limited by regulations or professional associations. The fundamental research system is separate from product development. Radical innovation is driven by public orders or by learning focused on capital goods (large public infrastructures); it requires large investments and a long timeline (for European countries including France).

- **Social-democrat system**: the idea is that of socialisation through institutions and negotiation (including economics and the consequences of innovation). Research and innovation are geared towards solving social and economic problems and depend on the availability of natural resources. Innovation concerns sectors that meet a social demand: health, safety, environment or use of natural resources. Training is driven by an egalitarian ideal.

• **Corporatist system**: large companies are the main vectors of innovation and skills development, based on solidarity and mobility within a large-sized and diversified conglomerate. Research and academic training are disconnected from applications and industry. Some innovations remain uncoded, that is, they are limited to a single enterprise. Education is general and homogeneous; specific skills are developed within the company. Innovation is about adapting products and procedures (profitable incremental innovations) before being about inventing new products. It can above all be seen in sectors requiring coordination between various skills: automotive, electronics and robotics industries. The system tends to be enclosed in its own national space (case of Japan).

The European Research and Teaching Space

The link between the sciences and society is still undergoing many transformations. We shall now examine the situation in Europe over the last 30 years.

Since the oil crisis of 1973 and the setting up of the European Community, public intervention in research matters has been headed by a social movement with the aim of developing 'useful science' instead of science that simply satisfies the curiosity of scientists. National and European research programmes have been created to 'encourage' researchers to refocus their work on the problems with which society is faced: research on alternative energies, protection of the environment, support of economic dynamics and job creation. Since the 1990s, it has not simply been a question of deciding which development to focus on, but rather one of how to set up overall 'governance' of the sciences, find the right technico-economic boosters and improve the performance of the research and higher education system. It has been a question of developing synergies and cooperation, cross-cutting approaches and networks, as well as centres of excellence and research or skills clusters.

Joint research in Europe has been marked by the setting up and re-negotiation of a series of politically orientated treaties (Box 1.4), whose influence on research can be felt. For example, the fair return principle defended by member countries has led to restrictions and to a dispersal of European Union (EU) effort. This in itself is something that is denounced and thwarted by more voluntaristic politicians.

In 1968, the common research centres (CRCs) employed 2,500 researchers, engineers and technicians. These centres were to go through a series of crises; their missions were redefined until, finally, at the start of the twenty-first century, they only held a marginal place in the European research system. Although the work undertaken in these centres, by EURATOM, led to some important results, the efforts of member countries did not form a consistent whole. In 1974, the idea was launched to gradually create a 'European research space'. This was to be based on four instruments: (i) the coordination of national policies via the Scientific and Technological Research Committee (CREST), which grouped

> **Box 1.4**   *50 years to build a European research space*
>
> 1952: Treaties on coal and steel (European Coal and Steel Community: ECSC) and on the atom (European Atomic Energy Authority: EURATOM). The latter was conceived as a common basis for development. Its history has had a strong impact on community-wide research.
>
> 1957: Signing of the Treaty of Rome, which does not mention research.
>
> 1974: First community-wide research actions.
>
> 1983: First R&D programme.
>
> 1986: Signing of the single act with an additional chapter devoted to research.
>
> 1993: Maastricht Treaty extending the scope of the EU's intervention in research matters.
>
> 1994: 4th framework programme (FP) on research and development, setting community research commitments at around €3 billion a year.
>
> 1997: Treaty of Amsterdam with simplified procedures for research.
>
> 2000: Declaration of Lisbon with a view to creating a European research space.

together the directors of national administrations; (ii) the creation of a European Science Foundation; (iii) the setting up of prospective studies to pave the way for European policies; and (iv) the grouping together of all R&D actions within a framework programme. With the Treaty of Rome making it possible to test new forms of action, the European Community Commission launched its first framework programme. Rather than carrying out domestic research in the CRCs, the existing research teams in the various countries united their efforts as part of this European programme. This type of action was further promoted in 1987 as part of the Single Act. Similarly, the assessment of programmes, introduced at the end of the 1970s, became institutionalised in 1987. Debates about the framework programme focused on the people involved in research that needed to be mobilised (that is, the role of industrialists, small and medium-sized enterprises (SMEs), individuals versus small teams and large consortia), the procedures to be implemented for selecting projects, assessing programmes, involving countries, sharing costs (between the Commission and its partners), programmes lasting several years (four then five years), and programmes based on themes (without open calls for tender or research team structural support as in France). Only specific forms of intervention prevailed: projects with shared costs, theme-based networks and demonstrations (first experimental achievements involving future users).

The theme-based approach of community-wide action was transformed. With the energy crisis in 1973, the Commission launched a research programme focusing on non-nuclear energies spurred by the fear of a shortage in supply. Along the same lines, research programmes on materials and on the environment, health and working conditions were set up. Industrialists limited their commitment to the harmonisation of technical standards. EURATOM recentred its actions in two areas: nuclear safety and fusion (with a large community research instrument, the Tokamak or Joint European Torus, installed in Great Britain). With the economic crisis at the turn of the 1970s and 1980s, European programmes began to adopt a more industrial outlook in favour of information and communication technologies. 'National champions' from industry contributed largely to preparing for the launch of the ESPRIT programme. The objective of such programmes was to 'strengthen the scientific and technological bases of European industry and foster competitiveness'. The economic goal was later to be included in the Maastricht Treaty; it was no longer a question of backing public policy but of supporting industrial research. New programmes focused on materials, communication, industrial technologies and biotechnologies (the idea being to have an open European laboratory with no specific location, organised so that a specific problem was tackled over a set period of time).

Furthermore, it having been found useful to consult the stakeholders of the ESPRIT project upstream, this method was applied to new FPs. In addition to the role played by the European parliament, official bodies were set up to organise the consultation process: the Industrial Research and Development Advisory Council (IRDAC) and the European Science and Technology Assembly (ESTA). Another feature of these European programmes was the way they were disconnected from national industrial policies. They were not set up to support a given industry since they were 'pre-competitive' (involving upstream research or research conducted jointly by rival companies). Support for industrial policies was provided through intergovernmental agreements and the EUREKA system.

In the 1990s, the importance of energy issues began to wane whereas that of the environment, life sciences and food safety gained momentum. The share of the budget awarded to industrial concerns settled at roughly 60 per cent. However, the building of the European research fabric, academic networks, mobility and training remained marginal preoccupations. With the 5th framework programme, programmes were structured around several far-reaching 'key actions' (homing in on social issues). These were accompanied by more general research work and support for access to member state research infrastructures. They involved all necessary public and private research skills together with the actors concerned (industrialists, hospitals, territorial authorities, infrastructure management bodies and the public service sector), including the future operators. These key actions aimed to solve a problem rather than develop a specific type of research. This meant that applications moved from being the usual industrial innovations, in terms of products and processes, to innovations geared more towards services: telemedicine, e-commerce and so on.

Furthermore, since the Bologna and Prague agreements, national teaching systems had been engaged in a vast movement of reform and harmonisation while the EU Commission strove to create a 'European research space'. The important thing was to reduce the 'dispersion of national scientific structures', ensure the coordination of skills centres (large research infrastructures) and contain the brain drain while encouraging researcher mobility. There was a dual challenge: to globalise the economy and to give birth to a new knowledge society. An objective was defined in Lisbon in March 2000: each member state was to devote 3 per cent of its GDP to research by 2010, two-thirds of which was to be financed by the private sector. The idea was to foster the best way to overlap public research and industrial development. These objectives were linked to the economic and industrial competition taking place in Europe, the United States and a number of Asian countries. Given the competitive climate, knowledge produced through R&D was seen to be a competitive resource: science was no longer an end in itself but an economic commodity and growth factor. This did not mean that fundamental research was called into question – European production in this field exceeded that of the United States – but it was unclear what type of place and role this type of research should be given. The sticking point was how to transform the base of fundamental knowledge into innovations.

The EU Commission then undertook the following operations: (i) harmonisation of regulations to encourage the free movement of researchers and their work (action focusing on scientific careers, social protection and intellectual property); and (ii) structuring of the European space. The idea here was to ensure that national actions formed a consistent whole while remaining specific. The French research system, for example, was especially stable. Very few institutions were closed down. On the other hand, new organisations were created (for example, the National Research Agency), and contracting and coordination methods put in place. In Germany, scientific institutions continued to be largely autonomous in spite of the power of the research ministry. In Switzerland and the Netherlands, research policies were negotiated at length between the representatives of economic, scientific and social interests.

Research institutions (CNRS, universities, agencies and so on) were also caught up in a campaign for social redefinition, and even social restructuring. The 6th framework programme introduced national programme networking and anticipation of the scientific and technological needs of other EU policies and research infrastructures (for example, the GALILEO satellite). The EU also reviewed financing models. Alongside projects with shared costs and theme-based networks, integrated projects and networks of excellence sprang up. With integrated projects, problem-solving was delegated to a consortium while the networks of excellence strove to prevent emerging fields from being fragmented and speed up new knowledge production rates. Nevertheless, some considered that all of these efforts were not enough: the financial resources remained marginal in terms of the overall European budget and compared with the spending to which the EU countries agreed. There was also a suggestion to reinforce the European Science Foundation to give it the kind of weight enjoyed by its American counterpart the National Science Foundation (NSF).

Conclusion: Assessing the Transformations of the Early Twenty-First Century

This chapter has traced the emergence of the sciences, how they became autonomous and the ways in which they were structured and linked with society. The situation has changed substantially over time, and continues to do so before our very eyes. It is too soon to measure the extent of changes to the institutional frameworks of science. Social science research has yet to report on the possible effects of such transformations. There are many changes involved: changes in discourse, perception, values and the very conception of science and the way it is steered, changes to institutional frameworks, organisations, research practices, direction, content and so on. The history and the sociology of scientific institutions provide various analysis frameworks and approaches that can be applied in an effort to understand the dynamics at work at the start of this next century. It would also be helpful to have new analysis frameworks.

Furthermore, there is no certainty as to whether a social consensus about the new directions in which the sciences are heading actually exists. There are many questions about the role and place of fundamental research and public research, about the status of researchers and how to assess them, about the actors to be involved in the system, about the role to be adopted by the 'scientific community' itself and about how all this new knowledge produced is to be filtered down through society. The social powers driving the world of science are many and varied and do not necessarily converge, depending on the countries, disciplines, institutions and levels (local, national and European). An analysis of all of these aspects must be able to rise to the challenge of building an overall assessment of the movement without losing sight of the diversity produced here and there.

Notes

1 This theory was criticised by some historians who pointed out that the same values could be found in non-Puritan Protestants and in Catholics, while others did not even see these values reflected in radical Puritans (Freudenthal in Ben-David, 1971).
2 Within the framework of the International Council of Scientific Unions (ICSU), created in 1931 and following on from the International Research Council (IRC), set up in 1919 to support the military alliance.
3 The International Committee on Intellectual Co-operation (ICIC) was founded in 1922 by intellectuals engaged in unification and universal pacification. It was linked to the League of Nations.

Recommended Reading

Ben-David, J. (1971), *The Scientist's Role in Society: A Comparative Study*, Englewood Cliffs, NJ: Prentice-Hall.
Grossetti, M. and Mounier-Kuhn, P.E. (1995), 'Les débuts de l'informatique dans

les universités. Un moment de la différenciation géographique des pôles scientifiques français', *Revue Française de la Sociologie*, **36**, 295–324.

References

References in other chapters: Price (1963) in Chapter 5; Vinck (1992) in Chapter 7.

Amable, B., Barré, R. and Boyer R. (1997), 'Diversity, coherence and transformation of innovation systems', in R. Barré, M. Gibbons, J. Maddox, B. Martin and P. Papon (eds), *Science in Tomorrow's Europe*, Paris: Economica International, pp. 33–49.

Barber, B. and Hirsch, E. (eds) (1962), *The Sociology of Science*, New York: Free Press.

Bernal, J.D. (1939), *The Social Function of Science*, London: Routledge & Kegan Paul.

Grelon, A. (2001), 'Emergence and growth of the engineering profession in Europe in the 19th and early 20th centuries', in P. Goujon and B. Hériard Dubreuil (eds), *Technology and Ethics. A European Quest for Responsible Engineering*, Leuven: Peeters, pp. 75–9.

Grossetti, M. (1994), *Université et Territoire. Un système scientifique local, Toulouse et Midi-Pyrénées*, Toulouse: Presses Universitaires du Mirail.

Pestre, D. (1989), *Louis Néel, le magnétisme et Grenoble. Récit de la création d'un empire physicien dans la province française, 1940–1965*, Paris: CNRS.

Polanyi, M. (1958), *Personal Knowledge. Towards a Post-critical Philosophy*, London: Routledge & Kegan Paul.

2 The institution of science

Historical sociology shows that science is instituted as a relatively autonomous and disengaged social space, dedicated to the production of objective knowledge. It outlines the conditions behind its emergence. Now, we are faced with a peculiar phenomenon: an institution and organisations that sociologists are striving to describe in an effort to understand the way they work. From their research, a new form of analysis has arisen, sometimes qualified as the 'institutional sociology of science' or the 'sociology of scientists'. This sociology of science was in fact born with the work of the American functionalist sociologist Robert K. Merton (Box 2.1).

Box 2.1 *Merton and the founding of the sociology of science*

Merton began his work at Harvard University where he was influenced by science historian George Sarton, founder of the review *Isis* – which brought together scientists, philosophers, historians and sociologists – and the *Osiris* collection of monographs, associated with the review. Merton published his first works in the review. He joined with Pitrim A. Sorokin, a rural sociologist and head of the Harvard sociology department, whose lectures he had followed. He also spent a lot of time with the young Talcott Parsons, an instructor working for Sorokin who at the time was carrying out an analysis of the fluctuations in truth systems (conditioning by the dominant culture of what society considers as true or false). Sorokin showed that the practice of scientific method is the result of the spread in society of the primacy of sensuality versus faith in the formation of truth. He endeavoured to provide quantitative measurements of the phenomena studied. Merton followed suit, founding his research on a quantifiable documentary base: a biographic dictionary and a base of articles.

Merton was interested in the evolutionary forms of different institutional spheres, the interaction between scientific and economic development and the reciprocal adaptation of positive sciences and cultural values. He developed a theory according to which interests, motivations and social behaviours within an institutional sphere (economy, religion and so on) maintain relations of interdependence with other spheres (including science). This interdependence comes from the fact that individuals have different roles and social statuses, corresponding to the institutional spheres in which they move; these spheres are therefore only partially autonomous. He underlined the role of the *puritan ethos* of the seventeenth century in the institutionalisation of science. Merton's analyses follow on from those of Weber on the meaning of ascetic rationalism for the development of scientific empiricism. According to Merton, the transformation of scientific interests (that is, which problems should be studied) is linked to the dominating values and interests in society. He pursued this research in his PhD thesis in 1938 (Merton, 1938).

Later, he showed that science only develops in societies where there are specific tacit values. On the contrary, in Germany, in the 1930s, hostility with respect to science was amplified in two ways: (i) scientific methods or results were said to be contradictory to the fundamental values of society; and (ii) a sense of incompatibility was maintained between *scientific ethos* and the *ethos* of other institutions. As of 1933, both of these approaches converged to limit the scope of science leading to the exclusion of Jewish scientists for the benefit of Aryans alone, to the submission of research to the immediate needs of industry, to the dependence of researchers with respect to politics, to a general reign of anti-intellectualism and to the difficulty of applying a critical approach to the results of Nazi research.

Merton explained the notion of scientific ethos in his 1942 article entitled 'Science and technology in a democratic order'. In this article he presented his conception of the normative structure. In spite of the diversity of scientific disciplines, the cultural values that influence them form a common cultural reality. Science can be sociologically defined as an institution based on a set of values and norms with which the scientist is supposed to comply; it is neither a set of knowledge items nor a set of methods. Merton identified the norms of this institution and underlined that the sociologist's job is to study the conditions according to which these norms help to regulate scientific activity: study of the influence of institutional norms on researchers' behaviour.

After 1957, he worked on quarrels between scientists about priorities and identified new research themes: the origin and perception of multiple discoveries, the ambivalence of norms, the prestige and forms of cumulated advantages, the forms and functions of competition between researchers and the procedures for assessing scientific work.

He published over 50 articles in the sociology of science field – see complete bibliography published in *Social Studies of Sciences*, **34** (6), pp. 863–78, December 2004. He was awarded many prizes and tokens of recognition and was notably the first sociologist to be admitted to the National Academy of Sciences (NAS). He is one of the most often-cited sociologists, notably for his theoretical contributions to general sociology. His work influenced Eugène Garfield, creator of the Science Citation Index.

The Normative Structure of the Scientific Community

Merton paved the way for the sociological study of the sciences, whose regulation mechanisms he studied. He built a middle range theory (neither a general theory of society nor a local interpretation of limited phenomena), which reported on the workings of science as a separate and autonomous institution. This theory explains both the individual and collective behaviours that promote this institution. Thanks to this institution of science, scientific rationality can be practised and knowledge built up and disseminated in society; increasing knowledge is the goal set by society for this institution. The norms governing the behaviour of scientists make this goal achievable; they constitute the normative structure or ethos of science. It is this structure that fosters the advancement of objective knowledge,

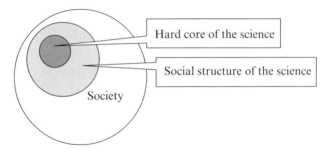

Figure 2.1   *Hard core and society*

protecting it from being trampled on by society, ideologies and specific interests. Merton thus lays the basis for the institutional analysis of the sciences by focusing on behavioural norms, social and professional habits and the values and ideas that guide the behaviour of scientists. He in particular studies the social workings of science, the process used to check scientific theory, which is fundamentally social given its public nature. Merton looks into the conditions that enable neutral and objective science to develop. He drops any sociological pretension to explain the actual content of scientific activity: the hard core (Figure 2.1).

Behavioural Norms and Ethos of Science

In his 1942 article (reproduced in Merton, 1973), he focuses on the normative foundation of the scientific community. He makes a distinction between two types of norms: *ethical norms*, relating to professional behaviour, and *technical norms*, relating to cognitive aspects (the logical and methodological rules of science). These two normative registers are interdependent. The advancement of knowledge depends on the implementation of technical imperatives (having empirical and reliable proof, ensuring that there is logical consistency) and particular methods (specific to each discipline). These technical norms depend on ethical norms that help to reinforce their performance and provide them with a moral guarantee. Conversely, ethical norms are the result of both the end purpose of the social institution of science and its technological imperatives. Moral imperatives have a methodological *raison d'être*. Merton says that methodological specifications are often both technological solutions and moral obligations. They are specifications that are just as moral as they are methodological. They are closely linked and indebted to the logical rules of science rather than to the form of society in which they emerge.

According to Merton, sociologists should study the former while the latter should be left to the care of epistemologists. He studies the ethical rules and describes the universal morals guiding all scientists, that is, the ethos of science: 'that emotionally toned complex of values and norms which is held to be binding on the man of science' (Merton, 1973, p. 268).

These norms are legitimate preferences and specifications, linked to the values of the institution. They are not encoded, but interiorised by scientists.

To underline the existence of these norms, Merton refers to various scientific writings and the indignation caused when the norms are not complied with. He defines four ethical norms, or institutional imperatives:

- **Universalism**: scientific productions and the awarding of tokens of recognition to researchers are subject to pre-established impersonal criteria. Universalism is the opposite of particularism, which focuses on personal criteria or belonging (gender, social belonging, national origin and so on). This norm is implemented through review mechanisms that are based either on public and transparent deliberation between specialists, or on 'double-blind' mechanisms in which the author of a text is not revealed to those in charge of the review (referees) and whose names likewise remain anonymous for the author. In the instructions given to those in charge of assessing an American scientific review, the following can be seen for example: 'Do not provide information in your review that reveals your identity and do not seek to discover the identity of the authors'.[1] Merton associates this instruction with the meritocracy ideal that prevails in science and which pleads in favour of scientific careers being open to appropriately skilled individuals. It is up to sociology to analyse the systems set up within the scientific community and their compliance with such a norm, by describing, for example, the workings of review committees or recruitment juries.

- **Communalism** (or communism): discoveries constitute collective property, produced in a collaborative manner and destined to advance society as a whole. This norm is the opposite of individual appropriation and secrecy; it requires researchers to communicate their results rather than keeping them for their exclusive use alone. This norm is implemented through the setting up of publication systems (annals of scientific societies, scientific reviews, lectures in conferences, online pre-publication and electronic reviews ensuring that results are quickly and broadly disseminated).

- **Disinterestedness**: scientific productions are public and verifiable. Scientists must report on them before their peers, which means that they are encouraged to seek the truth, produce reproducible results (technical norm) and publicly unmask any erroneous theories, data of poor quality (biased or falsified) and those having produced such data. They are supposed to strive towards the advancement of knowledge and not towards the advancement of their personal interests or those of a specific group, whether financial, ideological or professional. This norm is linked to the values of altruism and integrity.

- **Organised scepticism**: researchers and scientific productions should be systematically assessed using empirical and logical criteria that are not attached to any specific belief. This prevents statements that have not been thoroughly examined in the light of technical standards from being accepted prematurely. It requires researchers to remain open to rational criticism of their work and that of their colleagues. This means that scientists' work is systematically submitted to the critical assessment of colleagues, who sit on review boards. It also means that these colleagues are expected to furnish a considerable amount of anonymous and unpaid work as they examine and constructively comment on the texts submitted for review. In the instructions given to the assessors of an American review, the following instruction can be found: 'One of the greatest services that reviewers perform is the development of the research of members who submit their work. It is critical that *all* work submitted, regardless of whether or not it is accepted for the program, be improved by the feedback garnered from the review process. . . . Always maintain a professional, polite tone to your review. Authors deserve to be treated with respect, regardless of your evaluation of their work. . . . be sure that your comments are directed toward the ideas in the manuscript and not toward the authors. Be open-minded to different theoretical framings. . . . Try to judge manuscripts based on how well they stimulate thinking and discussion'.[2]

Merton and his disciples were to complete this normative structure with norms on originality, humility, rationality, emotional neutrality and individualism. We shall come back to the norms relating to originality and humility in the context of disputes about priorities.

Transmission and Transgression of the Norms

The sociology of science also examines the way in which norms are passed down and the mechanisms that encourage scientists to comply with them rather than transgress them. Because researchers are not forced to either adopt or comply with them, what exactly prevents these norms from being nothing more but pious intentions?

The transmission of norms is different according to whether they relate to technical matters or morals. Technical norms are explicitly taught: they can be found in methods manuals. The transmission of social norms, on the other hand, is by and large implicit. They are learnt from contact with other scientists, their morals and their habits, during the socialisation process in which young researchers identify with a group of scientists to which they would like to belong. They are handed down through the examples set by senior members of the research world, through the precepts outlined during scientific activity. Once interiorised, these norms fashion scientists' professional conscience and behaviour, to the extent that they become distinctive traits of their personality.

Scientists can be tempted to transgress these norms. They are, after all, subject to pressure from competition and may be tempted to use illegal means in order to oust a rival. However, such behaviour is negligible, according to Merton, and, in the long term, it does not work as it leads to the scientist in question being sanctioned. Scientists comply with norms spontaneously and because they are encouraged or obliged to. Merton completes his analysis of the institution of science with a description of the reward system that backs up the normative structure: a system of social control. Norms are institutionalised because they are associated with the dishing out of rewards or sanctions.

Designed to encourage compliance with norms, these rewards are symbolic. They take the form of honorary prizes such as the Nobel Prize or grants for studying, travelling or research. A reward can also be the assigning of an eponym associated with a theory (the Mendel laws, the Bernoulli principle, the Gödel theorem and so on), a value (the Avogadro number), a unit of measure (volt, ampere, ohm, angstrom), an object (Halley's comet), a representation (Bohr's atom), or recognition of a scientist as 'the father of . . .' (chemistry for Lavoisier, sociology for Comte). Rewards can also take the form of nominations (honorary member of an association, a working committee, a scientific committee, an editorial board and so on), or honorary titles, positions (in research, education or management) or missions. Having one's text accepted for publication in a review or by a scientific editor, receiving a favourable opinion from one's colleagues, being invited to speak at a conference, being cited in publications, in a manual or by a historian and, more generally, being accepted and recognised by one's peers, are all forms of reward that encourage researchers to comply with the norms governing behaviour in their community.

Researchers are thus supposed to comply with the norms of their institution and submit themselves to the social control operated by their peers (and not by some kind of hierarchy). They are encouraged to contribute to the advancement of knowledge via an evaluation of their work based on criteria that are independent of their personal qualities (it is only the quality of their work that counts). This evaluation is performed by review committees (for texts to be accepted for publication), scientific committees (for the allocation of research subsidies and positions) and juries (for the assigning of titles, ranks and scientific prizes).

Endowed with technical and ethical norms and a system of rewards ensuring the social control and compliance of behaviour with norms, Merton's scientific institution is a model of democracy: scientists are impartial in their judgements (both open and critical). They exercise self-control and mutual control over one another – with the youngest also being called on to evaluate senior members – without there being a need to institute some kind of form of superior authority, a state with authority to legislate, a police force or legal system that ensures laws are followed. The scientific institution is thus a model of democracy for the whole of society (Figure 2.2). Its development is furthered by the fact that the society surrounding it is itself democratic. It is far removed from the monarchic systems of the past.

Priority Disputes and Scientists' Ambivalence

Sociological investigation nevertheless reveals that scientists do not always come together as colleagues and that their meetings are often the scene of considerable conflicts. As they pursue their selfless quest for the truth, rivalries are expressed, controversies break out and competition sets in. Scientists' claims are not always appraised in a cordial manner or in the light of impersonal, disinterested criteria alone. The quest for the truth sometimes gives way to a quest for personal recognition.

In priority disputes, for example, the goal is to determine who should get credit for a given discovery from among several scientists claiming it to be their own. The discussions here no longer take the form of disinterested assessments;

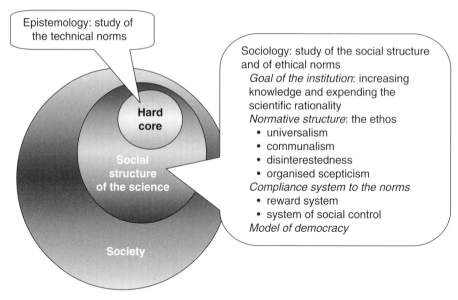

Figure 2.2   *Social structure of science*

they are about the social recognition that should be given to one individual and refused to others. To obtain this recognition, scientists sometimes adopt behaviour that is not very compliant with the norms: accusing others of fraud and plagiarism, stooping to libel, insults and defamation or even minimising others' contribution. There are many examples of this, not only from among the famous scientists and models of the past but from among contemporary scientists too. Isaac Newton claimed priority over Gottfried Leibniz with respect to the invention of differential calculus. When he presided over the Royal Society, he set up a committee in charge of deciding who should receive credit for this discovery, but he was also careful to place at the centre of this committee scientists in whom he trusted and whose activities he directed, going even so far as to anonymously draft the preface to their report. This was not the only quarrel about priority in which he was involved; according to Merton, he wrote around 20 texts in which he claimed he should take priority for a number of discoveries.

Newton is no exception. Similar quarrels have opposed René Descartes and Blaise Pascal, Galileo Galilei and several of his contemporaries; the Bernoulli brothers, father and sons (to the point that Jean threw his son Daniel out of the family home when Daniel was awarded a French Academy prize that he himself had wanted); John Couch Adams and Urbain Jean Joseph Le Verrier (about calculations about the position of Neptune); and Joseph Lister and Jean-François Lemaire (about antisepsis). More recent examples are Roger Guillemin and Andrew Schally, who were joint Nobel Prize winners (Wade, 1981), Robert Gallo and Luc Montagnier (about the AIDS virus), whose quarrel was settled by the American and French presidents, and, in 2004, the palaeontologist Michel Brunet and his geologist Alain Beauvilain (discoverer of Toumaï man). John

Flamsteed (astronomer to the king of England) said that Edmund Halley was a lazy, good-for-nothing thief who had stolen observational data that had been entrusted to Newton. These quarrels are often bitter and are part of the social relations between scientists. It is normal for scientists to quarrel, but where does this leave us in terms of the normative structure? Merton suggests getting around this difficulty by adding two other norms to the structure, those of originality and humility.

Priority disputes can be explained by the fact that many discoveries appear simultaneously but independently of each other. For example, when Wilhelm Röntgem discovered X-rays, other scientists were also aware of this phenomenon. Röntgem was simply the first to declare his discovery, which is why he went down in history while the others were forgotten. Merton cites the survey carried out by Ogburn and Thomas (1922), which draws up a list of 148 simultaneous discoveries. Convinced that this is a normal phenomenon, Merton writes: 'The pages of the history of science record thousands of similar discoveries having been made by scientists working independently of one another' (Merton, 1973, p. 371).

Nevertheless, a simultaneous discovery is not the only condition for a quarrel to break out. When Charles Darwin and Alfred Wallace discovered the theory of evolution at the same time, they shared credit for this discovery and did not quarrel. Similarly, before publishing his method for calculating variations, Leonhard Euler waited for his young colleague Joseph-Louis Lagrange to publish his. Merton refers to these examples as cases of *'noblesse oblige'*. They confirm that simultaneous discoveries do not necessarily lead to priority disputes.

Referring to psychological causes linked to human nature (egotism), or to the specific profile of scientists (their egoism and combativeness), is not of much use either when attempting to explain why quarrels break out. The first causes are too general to report on a phenomenon that is typical of the scientific institution while the second are contradicted by many upright men of science who may lack social ambition but who nevertheless get involved in quarrels about priority. James Watt and Henry Cavendish, known to be humble and disinterested, were involved in quarrels that set them against each other. So, one does not have to be egoistic, proud or ambitious to quarrel in the world of science. Furthermore, often the colleagues of these discoverers get directly involved in their quarrels without drawing any personal interest from such involvement other than making sure the truth is respected, according to Merton. They become engaged on behalf of their mistreated discoverer colleague and manifest their moral indignation, which is a sure indication that a social norm has been violated. Thus, for Merton, quarrels about priorities are responses 'to what are taken to be violations of the institutional norms of intellectual property' (ibid., p. 293).

One might ask which norm has actually been violated. The answer is the originality norm and the recognition associated with it. Recognition is based on who takes priority for the discovery and stems from scientists' devotion to the advancement of knowledge, which is the institution's ultimate value and goal. The originality norm urges researchers to produce hitherto unpublished knowledge. Being the institution's supreme value, originality is the basis for peer recognition. There is a structural exchange between the institution and the scientist: researchers offer up their discoveries to the institution, at the institution's request, and in exchange the institution bestows renown and esteem on them. The institution drives scientists to produce original knowledge that they can then publish as their own. As the content of the discovery belongs to the community, the only stake that scientists can actually claim is paternity for the discovery. This is why priority disputes are not about the content of discovered facts but about the fact of being discovered or not.

Recognition is awarded by the scientific community. However, to obtain recognition, scientists must make their rights known. Now, it is not always easy to determine what share of the new discovery belongs to each stakeholder. When Descartes claimed recognition for Pascal's famous experiment (the measurement of atmospheric pressure using a mercury-filled tube on top of the mountains in the Auvergne), he justified this by saying that he was the one to have suggested the idea. Who should get credit for the discovery? The person whose idea it was or the person who actually turned it into reality? What share of the originality comes from the idea suggested by Descartes and what comes from the design and production of the experiment carried out by Pascal?

Scientists' excessive combativeness, stemming from this originality norm and their claim for priority, is counterbalanced by another norm: the humility norm. This norm can be seen in the non-aggressive behaviour of Darwin and Wallace or Euler and Lagrange. Scientific humility is expressed through the recognition of debt towards the work of one's predecessors or colleagues and through the tribute paid to them. This is summed up by Newton when he said, 'If I have seen farther it is by standing on the shoulders of giants' (letter from Newton to Robert Hooke, dated February 5, 1676).

Researchers are thus used to citing, and also thanking, the authors they use or who have inspired them. This humility is expressed in other ways too: scientists may publicly recognise the limits of their own work, underestimate their contributions, or spout a stream of carefully phrased remarks. This can be seen in the work of a philosopher describing the thoughts of his famous patron: 'all that is interesting and original comes from him. If there are any mistakes, they are mine' (on the front page of the PhD thesis in theology of Jean-François Malherbe, 1983).

The norms of originality and humility create tension within the social structure of science. They are at once contradictory (one promotes combat while the other promotes reserve) and complementary (researchers recognise that they have done little to advance science, but what little they have done is theirs). Merton refers to their 'ambivalence': while demanding to see their merits recognised, scientists profess to be relatively indifferent to questions of priority. This ambivalence stems from the institution's system of values and leads to a variety of behaviours: quarrels about priority and examples of *noblesse oblige*. This tension results in dynamics that are specific to science; it encourages researchers to seek originality, but discourages them from wanting to obtain it at all costs.

Limits and Validity of Norms

Many debates in the sociology of science have focused on analysing this normative structure. This has resulted in a much richer analysis but has also led to a
querying of its usefulness in terms of understanding the dynamics of science. The
main question is whether the norms defined by Merton really reflect the scientific
institution and the way it runs? What exactly is the scope of the scientific ethos
he describes?

Specificity and Universality

Is scientific ethos a necessary and sufficient condition for the development of
science? The norms of scientific ethos can be found in other social groups; they
are not specific to science (Stehr, 1976). According to Merton, however, the specific nature of science calls for a combination of several norms (the normative
structure), not just one norm in particular.

What is the degree of universality and stability of scientific ethos? Can its
analysis be applied to all scientific practices, over time, from one nation, organisation and discipline to another? Is it not a historically contingent state of fact?
Barnes and Dolby (1970) reproach Merton for not taking into account the variety
of disciplinary and historic situations. Each period has its own norms: (i) the
scientific amateurism of the seventeenth and eighteenth centuries; (ii) the autonomous and professional academic science of the nineteenth century through to
the first part of the twentieth century, which is the period corresponding to the
norms outlined by Merton; and (iii) 'Big Science', together with its heavy machinery, team work and international programmes. Furthermore, since the 1980s, the
period which perhaps marks the beginning of the fourth era, researchers have
been involved in the commercialisation of their scientific production. Perhaps
the normative structure varies according to the organisational and technical
conditions specific to the scientific activity.

Researchers do not form a single and homogeneous scientific community
to which the normative structure of science can be applied across the board. On
the contrary, there are local cultures whose members have specific behavioural
obligations. For chemists, the scientific role to be played in an industrial environment (Stein, 1962) corresponds to imperatives that are different from the
general scientific ethos. Stein talks of 'limited communalism' because, in their
professional role, scientists have to limit the sharing of information and reinforce
industrial property. Furthermore, researchers are not entirely free to define their
research subject; they have to bring their work in line with their employer's needs.
However, out of the two normative systems of industry and science, industrial
researchers prefer the second (Kornhauser, 1963). For science students, scientific
ethos hardly appears to be universal anymore.

In science, there is neither legislation nor an authority that checks behavioural compliance. The review authorities (commissions and committees) vary
in terms of their objectives, criteria and workings. Their strategies refer to local

and specific normative systems. Moreover, they devote a lot of time to defining their own assessment rules and criteria. Instead of universal moral standards in science, what actually exists is a series of local moral codes. Lemaine et al. (1969) suggest taking into account local 'relevant scientific communities' and studying the system of norms and rewards specific to each one, in order to understand researchers' behaviour.

Deviance

Merton identified accusations of fraud and plagiarism, examples of defamation and discredit with respect to competitors' work during quarrels over priority. Some 'off-norm' behaviour can be pointed out: theft of data, falsification of data, non-disclosure or delayed disclosure of results, and even communication about the results of experiments before these have actually been carried out. Merton suggests that such deviant behaviour sparks indignation, proving that a norm has been deliberately breached, unless this is the result of tension within the normative structure. However, what is the real extent of such 'deviant' behaviour: is it marginal or common practice?

A difficult norm to put into practice: organised scepticism

Norms are not always implemented owing to the practical difficulties with which scientists are faced when attempting to conform. When two theories about how to interpret a phenomenon are contradictory, the norm of organised scepticism should lead researchers to organise a crucial experiment, in order to decide which theory applies. In fact, researchers have to face just as many technical difficulties when attempting to reproduce an experiment as disagreements about the way the experiment should be performed (Storer, 1966). This argument is explored further by Thomas Kuhn, David Bloor, Harry Collins and Andrew Pickering.

Scientific conservatism

Bernard Barber (1961) observes different forms of scientific conservatism according to the scientific environment and methodology used. Researchers sometimes put up resistance to certain data or discoveries. The history of science is full of examples of resistant behaviour. This can be due to a discovery's incompatibility with solidly established models in which researchers trust (argument furthered by Kuhn), or to religious inconceivability, or the upsetting of social or political commitments or to the fact that the authors of a discovery lack legitimacy – for example, because they are a priest or an amateur and therefore unworthy of confidence. Barber shows that scientists do not always have the open-mindedness suggested by the norms.

Secrecy

The practice of secrecy is contrary to the norm of communalism. Nevertheless, the history of science is peppered with famous examples of scientists jealously keeping their results secret. For instance, Flamsteed reproaches Newton for sharing his results with Halley, although Flamsteed had specifically asked Newton to keep the information secret. In other words, Flansteed reproaches Newton for complying with the norm of communalism.

The practice of secrecy can take on many forms. Descartes hesitates about letting Thomas Hobbes know about his work in order to prevent Hobbes from taking credit for it. Galileo informs the ambassador of Tuscany in Prague about his discovery of the planet Saturn using an anagram (SNAUSNRNUKNEOIETAKEY-NUBYBEBYGTTAYRUAS) whose meaning he explains several months later. Today, researchers announce partial results during press conferences and seminars in order to gain time, owing to scientific publication deadlines. They only deliver up a few details to underline that they take priority in the matter. This is either to give themselves time to check that the results are valid or to make the most of their competitive edge over others. Future Nobel Prize winner Guillemin wrote to his colleague to tell him not to reveal the Sephadex technique, which they had just discovered, during a symposium in 1961: 'The Sephadex story should be, in my opinion, kept for the Federation 1962 meeting in Atlantic City. We need time to investigate this striking separation' (Wade, 1981, p. 77).

The practice of secrecy also concerns relations with organisations in charge of assessing research projects and allocating subsidies to researchers. To prevent a promising subject or method that works from being revealed, researchers put forward research that has already been done, without revealing their genuine intentions. It should also be pointed out that the people sitting on review boards are often colleagues who may decide to use the proposals in their own interests. These review peers are therefore competitors who may just use the information gleaned from their evaluation work in their own research. Competition is often so ferocious that it leads to data being stolen. James Watson (1968) (discovery of the structure of DNA) recounts how, together with Francis Crick, they 'helped themselves' to the unpublished data of Rosalind Franklin and Maurice Wilkins.

Furthermore, owing to commitments undertaken with industrial and military partners (security imperatives), part of scientific productions may be kept secret or protected by patents while many results are only disseminated in small circles ('grey literature'). This can slow down the advancement of knowledge as it means that peers cannot use it in their own work.

The scientific institution is better at rewarding originality (openly gratified) than humility (leading to a certain silence). The norm of originality is thus awarded higher institutional value.

Attachment to ideas

Far from being disinterested, scientists are keen to defend their ideas and results from attack by their adversaries. They examine the arguments that are set against them in order to pinpoint their shortcomings. They do not easily let their results be called into question. Sometimes this attitude goes as far as a refusal to abandon a stance even when the arguments and empirical evidence put forward prove the scientist to be wrong. The French scientist René Blondot and his colleagues continued to observe emissions of N-rays when in fact the American physicist Robert

Williams Wood had secretly interfered with their equipment during his visit. Once Wood had published the report of his visit (his clandestine interference and the fact that this had had no effect on the convictions of the Nancy research group), and informed the whole world of the trick he had played, Blondot continued to receive firm support from the most eminent French physicists (including Marcelin Berthelot and Henri Poincaré), while his colleagues backed up his observations. Other French scientists refused to give any weight to the objections from abroad, arguing that they were due to frustrations linked to failed attempts to reproduce Blondot's results (Nye, 1986).

Scientists are often very involved in their work. They identify with it to such a point that some become embroiled in spectacular fraudulent activity, not in an attempt to obtain recognition from their peers but in order to defend their theoretical convictions. Work on heredity is often subject to fraud arising from conviction. Hence, an American zoologist discovered that the toads of the Viennese zoologist Paul Kammerer had been injected with Indian ink in order to demonstrate the hereditary transmission of acquired characteristics. In 1979, the famous English psychologist Cyril Burt, who died in 1971, was denounced for having invented the identical twins, separated and brought up in separate families, whom he said he had observed for years. He was also accused of inventing colleagues that nobody had ever heard of. He apparently also published 20 or so articles under different pseudonyms in a review for which he was in charge of the statistics column. The articles themselves confirmed or discussed his work. In fact, his results profoundly influenced the work of psychologists by underlining the importance of heredity in the transmission of intelligence. As a governmental advisor, he had also had a big influence on the design of the British school system. More recently, a German biochemist admitted that he had dispensed with actually performing the experiments he related because he was so sure of the results. He had published eight articles on the subject. An American immunologist was accused of having blackened the claw marks of mice with a felt-tip pen on his way up to his boss's apartment. He did this to convince his boss that his theories were justified. Other spectacular discoveries have also proved to be false and entirely man-made. In the Glozel affair, the archaeological site was set up using Gallo-Roman objects, attributed to the Neolithic period. In the Piltdown man affair, the skull had been tampered with to make it look old, and was in fact made up of a modern day man's skull and the jaws of an orang-utan. These revelations came late (the second came half a century after the discovery), and the authors of these fraudulent activities are, in some cases, still unknown.

Research assistants have also fixed data to avoid disappointing bosses convinced of theories yet unproved. There is a strong temptation to bend the results so that they correspond to a first impression or to correct an experiment in order to reproduce the expected results. It would seem therefore that there are many researchers who 'fiddle' with their models, instruments or data, or simply select the best pictures or data. Theories are launched on the basis of data that have been 'arranged' in a certain way, before they can be truly confirmed. Gregor Mendel apparently jump-started his data so that they corresponded to the 'right proportions' and Ptolemy 'calculated' some of his observation data, seen today as too precise given the instruments available at the time. Yet, these deviations from the norm seem to contribute to the advancement of knowledge.

Scientific Fraud

Fraud has always had its place in the annals of science. To support his theory of gravitation Newton used a corrective factor. Winner of the Nobel Prize for physics in 1923, Robert Millikan sorted his data into two groups – 'good' and 'bad' – in order to determine the charge of an electron. More recently:

- In March 1981, the review *Science* denounced the work of John Long, the specialist of Hodgkin's disease. He had made a name for himself growing the cells of this type of cancer. Cited by the leading lights of science in reference manuals – which had to be reprinted after purging any reference to the researcher – associated with the articles published by the team of Nobel Prize winner David Baltimore, and supported by subsidies awarded to him by his peers, Long was obliged to admit that the data supplied had been entirely fabricated and that he had falsified various results.

- In February 1983, *The Times* announced that the young and reputed American researcher from the Harvard School of Medicine, John Darsee, had been found guilty of falsifying results and fixing data. The School of Medicine was moreover condemned for not being very forthcoming in reporting the irregular behaviour of the researcher.

- In 1992, it was discovered that 20 years earlier, the Indian geologist Visham Jit Gupta had fabricated a bestiary of fossils from the Himalayas. He had disseminated false data in over 300 scientific publications, many of which had been co-signed by international palaeontologists.

- In December 2005, the work of the South Korean biologist Woo-Suk Hwang on human embryo cloning, published in *Science* and considered as revolutionary since May 2004, was suspected of fraud. After having acknowledged that he had violated the ethical rules of biomedical research by cloning his colleagues' ovules, Hwang was accused of having knowingly fabricated his experimental results. His laboratory was closed by the authorities. *Science* started proceedings to withdraw the incriminated article. The scientific community was shaken by this affair as it emphasised the weaknesses in the system used to validate the articles published in one of the most prestigious scientific reviews.

The few examples of fraud that have actually led to public denouncements (others are denounced behind closed doors and not let out) suggest that this is a more habitual practice than one might expect. In 1995, a Norwegian study declared that 22 per cent of scientists knew colleagues who had not complied with the norms. A file published by *Nature* (4 March 1999) showed that the proven cases of fraud have continued to increase over the years. *Nature* pleads in favour of stricter policing within scientific circles to ensure that codes of behaviour are better applied.

Why is there so much fraud? Several arguments are regularly put forward:

1   Fraud is possible because researchers do not sufficiently challenge the truthfulness of data produced by colleagues and presented in publications. Scientists tend to trust each other. Organised scepticism does not work. It is not easy to detect fraud. Such detection depends on the goodwill of researchers who volunteer to do review work. It also depends on the instrumental practices of each discipline, notably the difficulty of reproducing experiments.

2   Pressure to publish: publish or perish. The safest and fastest way to guarantee a career or obtain research credit consists in writing paper after paper. It would therefore appear that the harder the competition the more fertile the ground is for ethical shortcomings in the world of science. Researchers are subjected to so much pressure to publish that there is a growing number of cases where results are 'adjusted' or annoying details hidden under the carpet. Fraud can also be explained by the fact that the institution overvalues the norm of originality.

3   Financial pressures, very much felt in biomedical research, lead to results being dissimulated (case of Professor Ragnar Rylander whose research on the effects of smoke from cigarettes was financed by the tobacco industry). Privately funded research, accused of being the cause of increased fraudulent activity, has led to the idea that 'young researchers need to be helped so that they remain vigilant and are able to resist pressure'.

4   Scientific misconduct comes from young researchers. Not all researchers commit acts of fraud because they do not all hold the same positions; some, notably the youngest or those on the lower rungs of the social ladder, depend on the reputation that science can make for them if they are to achieve social recognition. The least advantaged suffer most from institutional pressure. This explains why it was in Robert Hooke's interest to get involved in quarrels about priority, while the noble and rich Cavendish had to be coaxed into defending his rights. There are researchers who disagree with this analysis, for whom fraud is committed by the generations wielding the power, those who no longer slave away in an effort to produce original data.

Proving that someone is in the wrong is a delicate business as with the case of the American biologist Baltimore, winner of the Nobel Prize for medicine in 1975. In 1991, he was obliged to step down from his office as Vice-Chancellor of the Rockefeller University in New York after supporting one of his colleagues accused of fraudulent activity in 1986. After being shunned for 10 years, both Baltimore and his colleague were reinstated by the American authorities in 1996 when the accusation was considered to be unfounded. Out of 1,000 cases dealt with by the American Office of Research Integrity between 1993 and 1997, only 76 actually led to the accused being condemned.

In reality, if the inventor of a stimulating theory makes some convenient alterations to the results or decides to present only the best results, this does not seem to be considered of importance. Moral indignation, as referred to by Merton, is often limited to a shrug of the shoulders and a pragmatic attitude: deceit, and its consequences for research, will end up being found out and put to right. As Roger Bacon said, 'truth is the daughter of time, not of authority'. In 1981, the president of the American NAS explained that fraud is almost always found out because science forms a system based on performance, democracy and self-control.

Yet, at the turn of the third millennium, scientific organisations are adopting a more cautious attitude. The trust that society had placed in science is crumbling. The scientific community will have to keep an eye open and make sure that ethical values in terms of appropriate conduct are handed down to the scientists of the future. Scientific organisations are busy drafting ethics charters, scientific quality standardisation committees are explaining and setting the standards behind best practices, and young researchers are receiving training and brochures entitled 'On Being a Scientist' published under the aegis of the NAS. In 1992, Denmark adopted a preventive approach by creating a committee to fight against scientific fraud. Germany published directives following the Hermann–Brach scandal (cancer specialists charged with falsifying their results, published in around 60 reviews). These directives and recommendations focus on the training of young researchers, the definition of responsibilities, the procedures for dealing with allegations of fraud and the tracking of laboratory activities in a way that prevents information from being falsified. When the *Nature* file was published in 1999, this also had an influence on several European scientific institutions in terms of the need to set up official control and monitoring systems. Research organisations are drafting best-practice guides and deontological research recommendations and setting up ethics committees. Ethics committees sometimes have to deal with suspected cases of fraud when these are not resolved within the laboratory. In Great Britain, the councils that finance research ask the researchers being subsidised to sign commitments according to which liability is transferred to them should there be any misconduct. Even scientific publishers are changing their practices with, for example, the creation of a Committee on Publication Ethics. It would seem that research organisations are gradually setting up more explicit regulations and procedures for prevention (in Europe) and control (in the United States), as though the ethos of science was falling short of the mark.

According to one legal expert:

It is important to remember that whatever professional autonomy is enjoyed by scientists, it is not a right but a privilege granted by society. If the public perceives the scientific community as rife with corruption, abusive of human or animal research subjects, or otherwise indifferent to the ethical requirements of research, then it will impose stricter mechanisms of accountability. (M.S. Frankel, *Proceeding of the Society for Experimental Biology and Medicine*, September 2000, 224 (4), 216–19)

What Do the Norms Really Represent?

Deviations in relation to the norms are so common and there are so many famous perpetrators that one is left wondering. Merton (1973) and Harriet Zuckerman (1984) say that there would be no norms if deviations to norms did not exist; the fact that a norm is overstepped does not prevent it from having a structuring influence. Furthermore, deviating from the norm does not afford any advantages in the long term. There is nevertheless plenty to question the ethos identified by Merton.

Barely Normative Norms

The norms of communalism and disinterestedness seem to be undermined by the practice of secrecy, fraud and attachment to ideas. Organised scepticism is questionable, even in cases where researchers acknowledge its worth. There are very few who actually take the time to check that published results can be reproduced. Researchers who publish critical comments are also a scarce breed while there are many who say that most publications are of poor quality. As for the norm of universalism, it has other weak points. For instance, reviews are influenced by a researcher's reputation. Indeed, laboratories make sure that they have people sitting on review committees or are known to their members. As for denouncing fraudulent activities, if this happens at all it is usually quite late in the day. Peers very rarely do the denouncing. This is left to rivals or assistants, shocked by the practices of their boss, or by journalists or fraud hunters. When a scientist is denounced by a colleague, this is usually because the colleague believes the theory to be of little credibility. The curator of the New York natural history museum tried to uncover the trickery of the Viennese scientist Kammerer. This was because, in Great Britain, the theory of the hereditary transmission of acquired characteristics was seen to be improbable. At the same time, however, the Soviets were in favour of it. A more in-depth analysis of acts of denouncement helps to understand how suspicions emerge, who starts the rumours and under what circumstances they are transformed or not into outright allegations of fraud.

Can it really be said that norms guide the behaviour of researchers and reflect the basic mechanisms of the scientific institution?

An Arsenal of Counter-norms

When questioning a series of scientists having worked on lunar rocks brought back from the Apollo missions, Ian Mitroff (1974) discovered that although they had good reason to conform to the norms outlined by Merton, they also had some excellent grounds for conforming to the counter-norms. They said that it was often more effective and usual to assess colleagues' productions by taking into account their personality, qualities, reputation and belonging than by scrutinising their data, concepts and theories or testing the results announced. The amount of trust bestowed in individuals influences how their work is evaluated. A

Table 2.1   *Norms and counter-norms*

Norms	Counter-norms
Belief in rationality	Role of irrationality (beliefs . . .)
Emotional neutrality	Emotional engagement and passion
Universalism: everybody is equal in the face of the truth	Particularism: some are a priori better than others
Individualism (against authority)	Social cohesion (against anarchy)
Communism (only the recognition is appropriated)	Private appropriation until the control of the use of the discovery
Disinterestedness	Commitment to defend his/her own interests
Impartiality regarding the consequences of the discovery	Feeling morally concerned by the consequences
Suspension of judgement; strict submission to the evidence	Capacity to judge from incomplete evidence
Validity related to protocols	Validity related to the author of the discovery
Loyalty only to the scientific community	Loyalty to humanity and to the general living of the human
Defence of the freedom of research	Rational management of resources

norm of particularism thus acts as a counterweight to the norm of universalism. In 1968, Merton himself published an article on the Matthew effect (see Chapter 5), where he exposes the hypothesis that the most eminent scientists tend to gain more prestige than other less eminent researchers for the same quality of work.

Similarly, the norm of private appropriation (solitariness) and secrecy, according to which initial results can be protected, counterbalances the norm of communism. Furthermore, secrecy stirs the curiosity and desire of colleagues, hence stimulating competition between researchers.

As for disinterestedness, Mitroff counters this with researchers' necessary attachment to their ideas and the need to serve their own interests. This behaviour provides the force necessary to complete projects, in spite of the threats raining down from all sides, including the threat of nature herself being reluctant to yield her secrets. Crick and Watson strove to demonstrate their idea that DNA had a helical structure:

'You have no proof', said Rosalind Franklin.
'We have faith', replied Jim Watson.

There are thus two sets of contradictory norms, it would seem (Table 2.1). However, Mitroff does not announce an alternative ethos of science. He shows instead that there is constant tension between these two sets of norms and that this is something that the scientific institution makes the most of. Like Merton with respect to the norms of originality and humility, Mitroff sets two sets of norms against each other in order to report on behaviour.

According to Mitroff, the norms identified by Merton are those of a handful of eminent scientists, idealised, generalised and raised to the status of institutional norms. They are institutional prescriptions, but they are only a handful of norms among other norms (local technical and moral norms) that scientists use. The dynamics of science depends on the fact that scientists are constantly commuting between both series of norms.

The Limits of the System

The normative structure described by Merton was at least clear and consistent. It seemed a plausible and fruitful way of characterising the institution of science; deviant behaviour provokes indignation, thus confirming the rule. The model is complicated with the introduction of two new, contradictory norms: originality and humility. They upset the explanatory system by adding tension and ambivalence. They refashion the system, which still has to explain the occurrence of a broader diversity of behaviours (fraud, disputes, demonstrations of modesty), but loses its ability to describe the specific behaviour of each scientist in a given situation: why does one put up a fight while another does not, why does one issue accusations in one case but not in another? Merton intimates that there are other factors entailed in the explanation, for example the position held by individuals within the institution (young scientists with no arrows to their bow or senior scientists already benefiting from a certain amount of esteem). Which provides the best explanation: the ethical norms or the position held within the institution? Other variables also come into play: the type of organisation and the norms it imposes on researchers (Stein, 1962; Kornhauser, 1963); the category of scientists and the variable importance they bestow on the different norms (Box and Cotgrove, 1968).

   Having seen so many examples of deviant practices, one might ask just how marginal or widespread deviancy actually is. If considered marginal, giving rise to indignation, it is proof that norms exist. If considered to be a general occurrence, provoking only superficial indignation, then it sheds doubt on the explanatory capacity of the normative structure. Explaining scientists' behaviour would involve juggling between two different sets of norms. Although this makes the model more realistic, it also does away with its wonderful simplicity and perhaps renders it less explanatory.

Ideological and Rhetorical Functions of Norms

This brings us to the question of the role of norms, whether they are universal or local, conventional or alternative. Was Merton wrong about the scope of their explanatory potential?

   In 1969, Michael Mulkay, inspired by the work of Kuhn, stated that these norms were nothing but ideals and values defended by scientists in their discourse. In practice, scientists actually have little interest in anything outside of their own field of predilection. From the moment they become socialised within a specific scientific speciality, what they are really aiming for is a cognitive consensus with their peers. They normally reject anything likely to upset the established models.

According to Cole (2004), Merton himself considered that the norms identified could not describe the workings of science, but were ideals regarding which scientists are ambivalent.

Norms apparently play less of a role directing scientific behaviour than legitimising it in the eyes of society (Barnes and Dolby, 1970). The ethos of science is thus a professional ideology that can be used to justify scientific autonomy. Researchers are quick to profess these norms but slow to actually put them into practice. They put them to use outside of their group in situations of justification or conflict.

Scientists choose norms because they serve their interests. They are just rhetorical resources (Mulkay, 1976). They help them to defend or legitimise positions or behaviours. They form a repository of moral rhetoric into which researchers can dip in order to defend themselves or call their rivals into question. Cambridge radio astronomers accused of practising secrecy, delaying the publication of results and holding up the advancement of science replied that they had to make sure their results were of top quality, adopt the measures necessary to prevent any incorrect interpretation and protect themselves to ensure that they took priority over a discovery on which their scientific recognition depended. Norms are thus rhetorical resources that are implemented when a practice needs to be justified. They are implemented during disputes to legitimise certain behaviours or condemn others. In the light of this, Merton's theory is less a model of the system of social regulation of scientists than an explanation of the internal justification discourse adopted by scientists themselves.

Norms are not so much descriptive as ideological. They were above all used as rhetorical tools when dealing with public authorities and society as a whole at a time when, at the end of the nineteenth century, scientists wanted to legitimise their practices and interests. In particular, the aim was to subsidise scientific enterprise while preventing society from having too much control over it. Similarly, at the end of the twentieth century, in a climate marked by a loss of confidence in the scientific institution, we are witnessing a renewal in the production of ethical norms. This time, however, they are accompanied by procedures, directives and bodies who are supposed to provide regulation and ensure social control over researchers.

Drawing his inspiration from the discourse (Box 2.2) of 'great' scientists, what Merton actually did without realising it was to reveal their ideology. This focuses on the autonomy of science and the eminently moral nature of men of science. Science being an eminently moral activity, it is assumed that any monitoring of the choice of research subjects, the methods used and the use of allocated funds is unnecessary. Norms are political resources whose role is to justify and legitimise the existence of the autonomous social structure. They reflect the ideal image of the scientist as promoter of the values advocated by American society at that time and still widespread among the American students of the 1970s. The scientific profession found that it was obliged to set up regular relations with the rest of society to ensure it had support and protection (Storer, 1966). Because science was not a service profession it had to win support differently.

Box 2.2 *Method – three ways to study norms: Merton, Mitroff and Mulkay*

Merton discovers the ethos of science by analysing a limited number of *texts* produced by famous scientists about their work and by studying morally indignant reactions to certain types of behaviour considered to be deviant. He takes selected statements from these texts and is reproached for not realising that they present an idealised vision of science in which the image of the scientist is at stake. The fact that these scientists refer to norms does not mean that these are the norms of the scientific institution. Following Merton, several studies (Gaston, 1978) investigate scientists' attempt to assess the extent to which they actually adhere to these norms.

Mitroff analyses a large quantity of first-hand documents (written by scientists) and *interviews* scientists.

Mulkay *analyses laboratory conversations*, notably when there is a controversy between two teams. He underlines the way in which the norms of scientific behaviour are used in a real situation. He reproaches both Merton and Mitroff for not having seen that texts and interviews produce standardised formulations used in a variety of ways according to the context.

The Mertonian Tradition

Merton left his mark on the history of the sociology of science by putting forward a model that gave rise to new research and many discussions. It is customary to associate Merton with a 'tradition', of which he is the founder. This tradition has been baptised 'traditional Mertonian sociology' or the 'institutional sociology of science'. After describing his model and the ensuing debates, we shall now briefly go back over the history of this research tradition.

Merton's sociology met with considerable success in the 1960s. His notion of scientific ethos is considered to speak for itself. Ben-David takes it up, saying even that simply expressing doubt about the autonomy of scientific research – supposing that scientific ethos is an ideological notion and that science is subject to the interests of a nation, a class and so on – is historically associated with Nazism, fascism and Stalinism. Conversely, adhering to the idea of science as an autonomous activity orientated by scientific ethos (which is what Merton and Polanyi thought) was perceived as a defence of democracy.

It was not until the start of the 1960s that Merton gathered several researchers around him. These were people who had been trained by him and drew their inspiration from his analyses. They worked on the social system of science and on the functional interactions within the scientific community. They postulated that this social system was rooted in a set of social norms and that the communication of knowledge between scientists was stimulated by the differentiated allocation of rewards, which themselves generated social inequality. Their research was based on empirical and quantitative surveys. They used, in particular, the Science

Citation Index (database listing all instances of publication citations in other publications), or equivalent databases to measure recognition and, indirectly, the quality of scientific work. Their subject of predilection was the normative foundation of the scientific community. They busied themselves studying the effective regulation processes and the way in which norms were instituted in these, for example, in the case of scientific editorial committees. Zuckerman and Merton (1971) identified the functions of such committees:

1   They guaranteed the scientific value of articles as they decided whether or not to authorise their publication. Because the articles represented the entire scientific community, the referees were committing the entire community when they accepted articles for publication.
2   They helped authors by taking on part of the results validation work.
3   They obliged authors to adopt a serious attitude to their research and only put forward articles whose conclusions had solid back-up and which complied with the state of the art. They urged authors to strive towards originality and offer a real contribution to the scientific community.

Three of Merton's disciples are particularly important:

• **Harriet Zuckerman** focuses on the scientific reward system, especially the conditions for awarding the Nobel Prize. She shows that the elite members of the scientific community have specific types of behaviour: they produce more and produce earlier than the average scientist: 3.9 articles a year instead of 1.4 for an ordinary researcher. The most famous – the scientific ultra-elite – tend to have more discussions among themselves than with ordinary researchers (Zuckerman, 1977). Nobel Prize winners have often studied with other Nobel Prize winners while the best researchers come out of laboratories that are already headed by famous scientists. A hierarchy is established inside the scientific institution and imposes strata among individuals, laboratories and universities.

• **The Cole brothers**, Jonathan and Stephen, examine the extent to which the reward distribution system complies with the standard of universalism (rewards assigned on the sole basis of the quality of scientific work); they conclude that this is not the case. In 1973, they publish a synthesis of their work on social stratification in science and the process behind social inequality within the scientific community. They focus on: the processes behind the assignment of positions within the social system; the relation between quantity and quality with respect to publications and recognition; the impact of the reward system on innovative minds; the influence of extra-scientific factors (sex, age, ethnic origin, religion) on the obtaining of recognition; and the nature of the relation between social stratification and scientific progress. Stephen Cole, who was Merton's assistant and colleague, put aspects of his master's theory to the test several times. In spite of the fact that some of the conclusions to his empirical anlayses invalidate Merton's theories, Merton does not revise them. According to Cole (1992, 2004), this is the case with the absence of the Matthew effect (see Chapter 5) and the influence of the degree of codification of discplines.[3]

• **Jerry Gaston** publishes several works on competition in science, in relation to the questions of originality and the practice of secrecy. In 1978, he publishes a comparative analysis of reward systems in the British and American scientific communities. He outlines the hypothesis that the way the reward system works is impacted by the organisation, financing and programming of research. He concludes that the more society gives free reign to decentralisation, the less the reward system is able to recognise orginal work.

A second circle of sociologists contributes to this tradition. It is made up of people from independent approaches, but whose results add to or complete those of Merton (Box 2.3).

Box 2.3   *The second circle*

Bernard Barber, who is close to Talcott Parsons, publishes work on the concrete modes of scientific discovery and the resistance of scientists with respect to these discoveries. He demonstrates the interest of the structuro-functionalist approach to the study of science. He also publishes one of the first collections of texts introducing the sociology of science.

Warren Hagstrom is concerned with the regulatory aspect of the sciences, but suggests that this regulation depends more on the exchange system than on the normative structure. For him, the desire for recognition is what basically motivates researchers. The institution uses this to achieve its goal.

Norman W. Storer publishes *The Social System of Science* in 1966. An admirer of Merton, he develops the Merton approach while at the same time directing it towards the sociology of professions. The autonomy of the scientific institution cannot be reduced to its independence with respect to society; it also depends on its internal organisation. Scientists accept the norms of their professional group because they need to maintain this social structure for their efforts to be recognised.

Diana Crane, a student of the historian J. Derek de Solla Price, studies the social circles in sciences, in particular the nature of communication and influence in the scientific field. She observes how scientific groups move towards and away from each other and explains this using the social norm of emotional neutrality. Researchers who are too emotionally attached to their own scientific ideas are frowned upon by their peers. Similarly, groups defending points of view that are too particular or exclusive, without sufficiently justifying these, are qualified as 'cliques' and considered as turning their back on the scientific spirit (Crane, 1969). Researchers are less attached to their group of belonging than to solving the problem they are studying.

Joseph Ben-David, after devoting his thesis to the social structure of professions, develops a historical approach to the sciences and universities, based on the notion of 'scientific role', institutionalised in society and associated with the autonomy of science idea. As of the 1970s, Ben-David is preoccupied with defending the notion of scientific ethos. For him, only this institutionalisation helps to explain that scientific activity has continued to exist over long periods of time, in spite of changes in paradigm. The scientific role is independent of scientific content. There is only one scientific community that envelops science as a whole.

Conclusion: A Revival of the Institutional Approach?

Merton engaged the sociology of science in a study of the way the institution of science worked. From this, he put forward a theory whose ingredients were the normative structure (scientific ethos) and reward systems. His work was further developed and discussed for almost 50 years. It resulted in a better understanding of institutional mechanisms and the effective functions of norms, at times more ideological and rhetorical than normative. At the end of the 1960s, scientific ethos as a descriptive notion was sharply criticised (Box 2.4). The idea that there was a single scientific ethos underlying all activity was called into question, notably using Kuhn's notion of paradigm. A new school of thought put the claims to universalism in science into perspective while the 'new sociologists' turned to the content of science, which had been excluded by Merton.

Box 2.4 *From one tradition to the next*
In 1974, during a conference of the American Sociological Association, a group of researchers envisaged the creation of a new, interdisciplinary, scientific society to study the social aspects of science. They asked Merton to be its president. The Cole brothers and Zuckerman tried to dissuade him from accepting. They suspected that the Society for Social Studies of Science along with its review, *Science Studies*, founded in 1969 by David Edge and Roy MacLeod, which then became *Social Studies of Science*, in 1974, would be highly critical of the Mertonian research programme. Merton actually became the first president indeed, in the 1980s, the new school of research began to gain in influence and became highly critical of the traditional sociology of science. Cole (2004) said that in the 1990s you had to be a constructivist if you wanted to be part of the social studies of science. Merton, who considered that most 'constructivist' writing was devoid of meaning, did not even deign to enter the debate.
In an article published in 1978, Ben-David attempted to give a sociological explanation of the different trajectories followed by the sociology of science in the United States (tradition arising from Merton) and in Great Britain (critical of this tradition). For institutional reasons, because the British science sociologists did not belong to sociology departments, their thinking about science was more philosophical, which explains their interest in the work of Thomas Kuhn, unlike their American colleagues who had had training in structural and functionalist approaches.

In the 1980s, the 'traditional' sociology of science, which the new research school had qualified as dominant, seemed to have disappeared. Its research programme had practically come to a halt. In the 1990s, few sociologists (Jonathan and Stephen Cole, Zuckerman and Thomas Gieryn, for example) continued to draw their inspiration from Merton.

At the beginning of the twenty-first century, the contribution of Merton and his disciples continues to be included in the academic training of the new generations of scientific sociologists. At the same time, the institutionalist analysis seems to be undergoing a revival. Between 1980 and 1990 in Germany (Peter Weingart (1982)), France (Gérard Lemaine, Terry Shinn) and the Netherlands (Loet Leydesdorff), research on the institutional dimension of science continued, without being based on the idea of a normative structure. In the United States, Henry Etzkowitz started work on research policies and the role of academic research in company dynamics. He took up again with the Merton tradition and identified a new normative structure of science incorporating the fact that researchers have been involved in the commercialisation of their scientific productions since the 1980s. When researchers have not yet acquired a reputation, such practices are seen in a negative light, from the standpoint of the old social norms of science. However, when researchers already have a reputation, this is strengthened by such practices. These researchers are then presented as models by their peers who admire what they do. Such attitudes reflect a change of norm in the academic milieu. They suggest that the institution of science is in the throes of a radical change: the switch from individual research to teamwork; the concern for efficient management of research activity and its organisation; assessment of the results of research with respect to both their commercial interest and their theoretical relevance. New social norms are emerging (Etzkowitz, 2004): capitalisation of knowledge and the norm of limited secrecy; interdependence between universities, the state and industry (the triangle model of Jorge Sábato (1975), the Leydesdorff triple helix model); autonomy of entrepreneurial universities; hybridisation of economic dynamics and the advancement of knowledge; reflexivity and the ongoing recomposition of university structures. For Etzkowitz, who analyses the entrepreneurial development of universities, including MIT, tension between norms fosters innovation. Innovation is above all the creation of new arrangements within institutional spheres that encourage technological innovation.

Notes

1 http://www.bpsdivision-at-aomconference.org/bpsreview/Guidelines.htm accessed 17 January 2005.
2 Ibid., accessed 17 January 2005.
3 Zuckerman and Merton (1971) suggest that the more codified a field is, the higher the level of consensus. Their demonstration is based on an analysis of the rejection rate for articles submitted to reviews: 80 per cent of articles submitted to the *Physical Review* are accepted while the rate is only 10 per cent for the *American Sociological Review*.

Recommended Reading

Barber, B. (1990), *Social Studies of Science*, New Brunswick, NJ: Transaction.
Cole, S. (1992), *Making Science: Between Nature and Society*, Cambridge, MA: Harvard University Press.

Crane, D. (1969), 'Social structure in a group of scientists: a test of the invisible college hypothesis', *American Sociological Review*, **34** (3), 335–52.

Merton, R. (1973), *The Sociology of Science: Theoretical and Empirical Investigations*, edited by N.W. Storer, Chicago, IL: University of Chicago Press.

Mitroff, I. (1974), *The Subjective Side of Science: A Philosophical Inquiry into the Psychology of the Apollo Moon Scientists*, Amsterdam: Elsevier.

Mulkay, M. (1976), 'Norms and ideology in science', *Social Science Information*, **15** (4), 637–56.

Storer, N. (1966), *The Social System of Science*, New York: Rinehart & Winston.

Wade, N. (1981), *The Nobel Duel: Two Scientists' 21-year Race to Win the World's Most Coveted Research Prize*, Garden City, NY: Anchor Press-Doubleday.

Watson, J. (1968), *The Double Helix: A Personal Account of the Discovery of the Structure of DNA*, New York: Atheneum (Norton Critical Editions, 1981).

Zuckerman, H. (1977), *Scientific Elite: Nobel Laureates in the United States*, New York: Free Press.

References

Barber, B. (1961), 'Resistance by scientists to scientific discovery', *Science*, **134** (3479), 596–602.

Barnes, B. and Dolby, R. (1970), 'The scientific ethos: a deviant viewpoint, *Archives Européennes de sociologie*, **XI** (1), 3–25.

Ben-David, J. (1978), 'Emergence of national traditions in the sociology of science: the United States and Great Britain', *Sociological Inquiry*, **48** (3–4), 197–218.

Box, S. and Cotgrove, S. (1968), 'The productivity of scientists in industrial research laboratories', *Sociology*, 2, 163–72.

Cole, S. (2004), 'Merton's contribution to the sociology of science', *Social Studies of Science*, **34** (6), 829–44.

Cole, S. and Cole, J. (1973), *Social Stratification in Science*, Chicago, IL: University of Chicago Press.

Etzkowitz, H. (2004), 'The evolution of the entrepreneurial university', *International Journal of Technology and Globalisation*, **1** (1), 64–77.

Gaston, J. (1978), *The Reward System in British and American Science*, New York: Wiley & Sons.

Kornhauser, W. (1963), *Scientists in Industry*, Berkeley, CA: University of California Press.

Lemaine, G., Matalon, B. and Provansal, B. (1969), 'La lutte pour la vie dans la cité scientifique', *Revue Française de Sociologie*, **10** (1), 139–65.

Merton, R. (1938), *Science, Technology and Society in Seventeenth Century England*, New York: Fertig.

Merton, R. (1942), 'Science and technology in a democratic order', *Journal of Legal and Political Sociology*, **1**, 15–26.

Mulkay, M. (1969), 'Some aspects of cultural growth in the natural science, *Social Research*, **36** (1), 22–52.

Nye, M. (1986), *Science in the Provinces: Scientific Communities and Provincial Leadership in France, 1860–1930*, Berkeley, CA: University of California Press.

Ogburn, W. and Thomas, D. (1922), 'Are inventions inevitable?', *Political Science Quarterly*, **34**, 83–98.

Sábato, J. (1975), *El pensamiento latinoamericano en la problemática ciencia-technología-desarrollo-dependencia*, Buenos Aires: Paidós.

Stehr, N. (1976), 'The ethos of science revisited: social and cognitive norms', *Sociological Inquiry*, **48**, 173–96.

Stein, M. (1962), 'Creativity and the scientist', in B. Barber and W. Hirsch (eds), *The Sociology of Science*, New York: Free Press, pp. 329–43.

Weingart, P. (1982), 'The scientific power elite: a chimera; the de-institutionalization and politicization of science', in N. Elias, H. Martins and R. Whitley (eds), *Scientific Establishment and Hierarchies*, Dordrecht: D. Reidel, pp. 71–87.

Zuckerman, H. (1984), 'Norms and deviant behavior in science', *Science, Technology and Human Values*, **9** (1), 7–13.

Zuckerman, H. and Merton, R. (1971), 'Patterns of evaluation in science: institutionalization, structure and functions of the referee system', *Minerva*, 9, 66–100.

3   The sciences as collectives

With Merton, science was studied as a unique, normative institution. The term 'scientific community' was coined in reference to the different parts making up science. However, Storer (1966) suggests that the normative structure does not go far enough to explain how these parts are integrated to form a whole. The internal organisation of science has to be taken into account too. This internal organisation projects an alternative image of science, as a set of different communities, which can be analysed in terms of profession or discipline. As we shall see, it is possible to go beyond these analyses by studying regimes of knowledge production.

The Profession

The scientific community can be seen as a series of specialised professions, each with its own internal organisation, in the same way as doctors, architects or lawyers belong to a specialised profession. To become autonomous, a profession has to have an internal organisation and be made up of members who are all eager to contribute to relations. In the case of medicine, there is a professional association and a formal deontology, but can the same be applied to scientific research?

Storer (ibid.) saw within scientific specialities professions, inside which there are well-ordered and regulated links between members. The ethos of science cannot fully explain how these links are actually regulated as it does not cover the way members of a group are attached to the group's norms. This attachment comes from the fact that researchers count on the upholding of their profession as a social structure within which their efforts are understood and recognised. Researchers play various roles within this structure: they make contributions as research novices, organise and manage the profession, are involved in debates about new collective directions, or develop group rules and relations with other groups. The profession is characterised by four aspects:

- It is **responsible for a body of specialised knowledge** that it upkeeps, passes on, extends and applies. In chemistry, for example, scientific societies organise the harmonisation of nomenclatures and manage the dissemination of compendia used as references for researchers, lecturers, experts and industrialists alike. In sociology, societies (for example, the International Sociological Association (ISA) or the European Association for the Study of Science and Technology (EASST)) organise working groups, research committees and theme groups where ongoing research is debated.

- It is **autonomous in terms of the recruitment, training and control** of its members. Entry into the profession of researcher or university lecturer is generally controlled by commissions made up of peers belonging to the discipline, and not managers of human resources. In French universities, disciplinary sections of the National University Council (CNU) and local 'specialist commissions' are in charge of this. Access to scientific societies sometimes depends on an existing member putting forward the names of possible future members.

- It sets up and maintains **regular relations with the rest of society** in order to gain support and protection. As the sciences are not service-orientated professions, they do not sell their expertise. In other words, their support comes mainly from teaching, subsidies and research contracts.

- It has its **own system of rewards to motivate and control its members**. Members are incited to follow the norms of their scientific profession because of the recognition they receive and their close relationships with colleagues during their years of training. This is an extended socialisation process. The sciences are social systems that motivate and control individuals through a system of mutual rewards granted to each other. When researchers receive recognition other than from their profession, for example, some form of personal remuneration, or are rewarded for sitting on a company board of directors, they may be tempted to sidestep the principles of their profession. The question of reward allocation is central to understanding the dynamics of this social system and the development of a body of knowledge.

Scientists are confronted with issues relating to professional recognition. However, the existence of scientific societies and training programmes does not entirely cover their need for recognition. Although chemists, lecturers, researchers and industrialists can be seen to make up an identifiable professional group, putting physicists into a single group proves to be more difficult. This is because physicists are associated with the academic world while at the same time competing with the world of engineers. Psychologists are also confronted with the difficulty of making sure their group is relatively closed; only those holding diplomas can set up as practitioners. As for economists, they are rarely recognised as a profession, outside of research and education; they have to think up strategies so that their speciality exists from both an academic and social point of view, as shown by Ashmore et al. (1989) in the case of health economists.

According to Storer, the sciences form simple social systems because their basic values are stable and there is no complex role differentiation. The main difference that might be pointed out is the distinction between junior and senior members.

The Diverse Social Roles of the Scientist

Within specialities, scientists differ according to the functions they take on (Znaniecki, 1965). An individual occupies multiple social roles or functions, simultaneously or over the course of their life: technicians (advisor, coordinator or expert), scholars, discoverers, systematisers, contributors, disseminators or fighters of truth. All of the roles fulfilled by an individual make up their social personality. The roles are linked to types of knowledge (technical, common sense, wisdom and sacred or absolute knowledge). These roles correspond neither to stable functions nor to statuses. For example, when there is rivalry between schools of thought – rational jousting against a backdrop of agreement about a set of shared cognitive elements – the dynamics of the debate can give birth to a specific fighter role.

The scientist's social personality depends on the complementarity between different roles, the balance of which varies throughout the course of his/her career, according to personal preferences and institutional contexts. Zuckerman and Merton (1972) distinguish between four roles:

- **Researcher**: functionally central, this role is linked to the development of knowledge. Its value is ostensibly promoted by scientists as it gives them the possibility of leaving their name to posterity. This role is sometimes subdivided into theoreticians and experimenters.

- **Teacher**: this role supposes that there is knowledge to be passed on so that an apprentice can be transformed into a member of the community. It reflects the socialisation process, which is more determining than education itself. Scientists often feel obliged to train their successors, but do not wish to spend too much time doing this.

- **Administrator**: this role concerns a variety of activities to do with management and scientific leadership, the search for partnerships and resources as well as labour organisation.

- **Regulator**: this role concerns the appraisal of work, individuals and teams, leading to the allocation of resources and recognition, including the authorisation to publish.

Professionals and Amateurs

Professional scientific groups have often been set up in opposition to other professional groups or amateurs. Such opposition helps to build the identity of scientific specialities and is relatively complex. Thus, when experimental research was being developed in biology laboratories, it appeared that adepts of naturalism stopped contributing to this new form of biology. In reality, works on the social history of the sciences show that amateurs have played an essential role in the building of professional university communities. They collaborate with established biologists in laboratories; some actually become academics. In fact, the notion of professional group tends to veil the diversity of amateurs and professionals within it. In the case of biology, professional biologists themselves have masked this diversity in so far as their identity is formed in opposition to a specific type of amateur.

The relations between professionals and amateurs are complex; amateurism can sometimes be found on the road to professionalism.

Scholarly Societies and Professional Organisations

Scientific societies, often specific to a discipline, encourage fundamental research, discussion of results and theoretical developments, and spread their knowledge via reviews, scientific publishers' collections or abstracts. They organise workshops where members have the opportunity to present their work, hence offering it up for discussion. The general atmosphere fosters debates about general theoretical questions rather than about specific phenomena. Little interest is shown in the practical knowledge linked to applications. This phenomenon can be observed in international conferences where the participants prefer plenary session discussions about major research trends rather than discussions on overly specific topics.

The dynamics of a speciality can also be explained by this type of relative preference. Automation, for example, arose from the production management problems facing industrialists. Researchers analysed concrete situations and designed models enabling local industrial improvements. During conferences, these models attracted greater interest than the concrete problems facing businesses; they raised intellectually interesting questions, some of which did not apply to the industrialists themselves. New intellectual challenges and research programmes came out of these discussions. Presented at different conferences, and explained in publications, models with a higher level of generality emerged and became the focus of attention. The researchers who were actually working on the specific industrial issues attached less interest.

Other professional societies are organised in order to encourage research and spread more applied knowledge. They bring together practitioners rather than researchers. These are, for example, associations specific to a sector (textile, agriculture, metallurgy, environmental management, sociology of intervention and so on). Although their members are practitioners, they encourage research, organise conferences and discuss feedback. They collect information, supervise discussions on knowledge provided in scientific publications, identify problems and finance research. They create incentives, such as honorary titles, prizes awarded for books, grants to finance publications and loans to carry out new experiments. They sanction those who do not get involved, who keep their inventions secret or whose work is not of good quality. These associations are often created to counter external adversity. This was the case with the iron and steel industry in the eighteenth century in Sweden where the aim was to promote research with master blacksmiths (Ben-David and Katz, 1982).

Scientific societies are places where scientific discussions take place, where mutual encouragement is given and knowledge is spread, but they also have a regulating, almost legislative and policing role. They influence the moral climate in which their members carry out research work. They constitute a reference from a normative point of view. Their members join up and discuss best practices and new analysis protocols. This is the case of clinical biology laboratories, for example, which benefit from feedback following the dispatch of anonymous samples. This then allows them to compare their work with that of others and undertake a general improvement process. Such scientific societies act as the keepers of rules, values and traditions. They provide their members with support that helps them to build up their identity. The members are thus encouraged to see themselves as actors of a collective adventure.

In a common declaration, the US National Science Academy, National Engineering Academy and Institute of Medicine announced: 'As members of the professional research community, we should strive to develop and uphold standards that are broader than those addressed by the governmental regulatory and legal framework for dealing with misconduct in science' (Alberts et al., 1994, p. 3479; NAS et al., 1992). Developed within scientific societies, these standards incarnate a form of collective consciousness of the professional group. They are part of the group's construction and provide its members with the opportunity to participate in the moral life of the profession, to test their professional ethics in relation to those of their colleagues, and to test the ethics of the profession in relation to society's expectations. These scientific societies play a regulating role while at the same time constituting relational and organisational resources for their members.

Disciplines

Scientists tend to form self-regulated communities in which individuals are relatively equal. Barber (1952) explains the multiplicity of these communities using the norm of individualism and points out that their members believe themselves incapable of judging related specialities.

The Emergence of Disciplines

In the Middle Ages, the subjects taught were progressively grouped into two categories: *trivium* (grammar, rhetoric and logic on which the art of reading was based) and *quadrivium* (including arithmetic, geometry, music and astronomy, and speculative tuition on numbers or harmonies). The universities brought together scholars and students. After much conflict with the ecclesiastical authorities, teachers obtained the right to decide which candidates should be awarded the *licencia docendi*, a certificate which then allowed them to go on to teach. The universities, whose members were men of the Church, set up an organisation allowing them to control recruitment in the same way as tradesmen decided which apprentices to take on. They created faculties within the universities. The first was devoted to the arts (notably foundation courses in language). Pupils entered at the age of 13 and graduated after six to eight years. This training led into three higher faculties (theology, law and medicine), where students could obtain a degree or become a doctor after about 15 years.

The arrival of new actors helped to regenerate the sciences, dividing them up and organising them into a hierarchy. New religious orders (Dominicans and Franciscans), more in contact with the world and engaged in heavy debates, notably with the Cathars, who stimulated their intellectual activity, turned towards the natural sciences. St Thomas Aquinas distinguished between two types of sciences: those stemming from known principles thanks to the natural enlightenment of the human intellect – arithmetic and geometry – and those arising from rules known only to those enlightened by a higher science – optics based on geometry and music based on arithmetic. Drawing inspiration from the Greco-Arabic models, these religious orders (for example, the Dominican naturalist Albert the Great and the Oxford Franciscans, including Roger Bacon), developed a scientific thinking that completed text commentary with original observations, experimentation and reasoning based on data acquired by the scholars themselves or by their correspondents.

Similarly, the development of trade led to the emergence of new fields of knowledge. Accounting and the calculation of insurance contracts, trade profit sharing among the members of a family group and their allies, as well as the calculation of exchange rates and letters of credit all required the skill of writing and calculation. The use of Arabic figures and the method of written calculation, making it possible to keep track of intermediate results, spread among astronomers and tradesmen. The emergence of paper, less costly than parchment, and the invention of new ways to present calculations explain the development of algebra (rules of 3, presentation based on equations, algebraic calculation). Arithmeticians made a name for themselves in trading towns. There were so many people living off calculation skills that they actually outnumbered the business world. New generations of scholars, humanists with training in the Greek language, dialectics and astronomy, assimilated these trade-based calculation techniques and adapted them to the study of movement, geography, the sky, architecture, optics and medicine. The conditions for the production of knowledge were transformed and new parallels appeared. With the arrival of printing (of both old and modern texts), observation reports and calculation results became more easily widespread in the West (Eisenstein, 1979). Hitherto remote data were compared and contrasted. New fields of knowledge became popular, to the detriment of others; some were regenerated as they assimilated new methods. The name 'human sciences' then referred to geometry and was opposed to the divine sciences.

In the sixteenth century, the first science academies were created. These sometimes granted salaries and created their own scholars' newsletter (*Journal des Savants*). In France, the structure of the academy was based on 'classes'. In 1780, there were six: Geometry, Astronomy, Mechanics, Anatomy, Chemistry and Botany. In 1785, Physics and Mineralogy were added to the list. Other academies were created in the provinces along the same lines. Science was carried out in these academies by individuals who shared and discussed their observations. The academies encouraged collaborative work, the setting up of commissions for a given issue and the organisation of competitions on a particular theme (for example, 'How to improve lighting in the streets of Paris', in 1764). They organised a systematic inventory of publications and highlighted new questions and contra-dictions. Although the research fields were organised according to the academy classes and the division into faculties, the scientific disciplines had not yet become part of the institution. On the other hand, with the Encyclopaedia, there was a movement to classify and structure the fields of knowledge (mechanics, analysis, astronomy, physics, chemistry and natural history) forming the identity of the disciplines. Denis Diderot and d'Alembert's tree of knowledge is still influential today.

Between 1665 (*Journal des Savants*) and 1829, 300 scientific periodicals were created across the world: Newsletters, Reports, Bulletins, Acts, Annals and so on. In 1800, the number of scientific reviews published amounted to roughly 100, while in 1850 the number had risen to around 1,000, that is, 10 times more.

It was not until the nineteenth century that an organisation based on dis-ciplines was instituted, with the creation of modern universities. In nineteenth-century Germany, the number of research schools kept growing. They claimed their autonomy (see Chapter 1) and tended to be specialised. Disciplines were created and structured as autonomous entities, like the nation-states. This dis-ciplinary conception of science in Germany was used as a model of scientific organisation that became widespread in other nations.

Disciplinary autonomy: chemistry

Throughout their history, chemists have pondered their identity, wondering what makes chemistry different from physics. Were they part of physics or not? This question is reflected in the Mendeleyev table, on which chemistry is founded. The table can be explained using the wave functions of quantum mechanics. Mendeleyev himself would have liked physics and chemistry to join forces under the umbrella of mechanics in the nineteenth century.

This question caused much debate in the science academies of the eighteenth and nineteenth centuries. For some, chemistry differed from physics because it worked on mixtures and complicated cases, situations that could not simply be reduced to a few basic principles of physics. It could have been classed as a branch of physics if it had been possible to push back the limits of calculation, but it was based on approximations stemming from experience rather than on theoretical intelligibility. The questions raised by chemistry came more from practice and craft-like activities than from theory. For others, on the other hand, there was only a qualitative leap between physics and chemistry in terms of the organisational complexity of matter. The calculations were instructive because they were based on models and concepts. Using diverging object constructions was therefore justi-fiable. Chemistry was also subjected to many caricatures. For example, chemistry required a long apprenticeship of the senses and body in order to develop the ability to read the clues. Chemical scientists had to be passionate about their art. Physicists, who calmly based their reasoning on deductions, saw chemists as mad scientists, eccentrics who were slaves to their passion, to the point that they ruined both their health and fortune.

Nevertheless, the *metier* of chemist underwent considerable transformations.

At the start of the nineteenth century, the new identity of the chemist was marked by the introduction of systematic and exhaustive studies, the standardisation of instruments and products, the drawing up of a nomenclature for substances, the definition of experimental protocols, as well as the ability to control and repro-duce experiments. The activity in laboratories became almost industrial. In the Liebig laboratories, a new generation of chemists were trained to use instruments and implement protocols in just four years. They came from across the entire planet to learn chemistry in Germany, flooding university and industrial labora-tories where they received systematic training. Chemistry was one of the first disciplines to become international.

Backed up by its contribution to industrial development and the design of products and processes, chemistry campaigned for its autonomy, claiming that it was able to raise fundamental questions and pursue its own research programmes. It became a positive science model, which could be compared neither with the deduction-based approach of old school physicists nor with the impassioned chem-ists of the previous century. Right up until the 1960s, it provided a model for scien-tific enterprise, taking part in industrial development and in the daily transformation of life thanks to the plethora of new products and materials that it produced.

However, chemistry faced a new identity crisis in the twentieth century with the growing prestige of atomic physics, the development of electricity and elec-tronics, and the calling into question of all sorts of pollution. Accused of causing many ills and perceived as boring, chemistry had to strive hard to be seen as an attractive science, a science that was concerned with environmental issues (new recycling materials and processes), and which encouraged the scientific inquisitive-ness of young people as new phenomena were studied (strange events, chaotic phe-nomena and so on). Associated with other disciplines (material physics, biology and biotechnology and so on), its frontiers continue to be constantly redefined.

The Birth of New Disciplines or Specialities

The emergence of the disciplinary approach to the sciences reflects a general gearing towards institutional dynamics. However, the specific history of each speciality calls for a more subtle and local analysis.

The emergence of a specific scientific role

Considering that the ideas needed to create a discipline have usually been around for some time, in different places, Ben-David and Collins (1966) suggest that it is above all important to look at the 'environmental mechanisms' that determine the structuring and institutionalisation of a new discipline. Their hypothesis is that a speciality develops in a given place at a given time because the scientists who are interested in it have the means of instituting a new intellectual identity and a new professional role. Studying the institutionalisation of scientific psychology in around 1870, they identified three categories of scientists (based on the number of publications), and traced the relationships of filiation between them from one country to another:

- **The precursors**: did not consider themselves as belonging to the new discipline and their pupils did not become specialists either. They were therefore not recognised by their contemporaries as members of the discipline although they were associated with it, often posthumously, by its 'historians'.

- **The founders**: trained disciples in the discipline without having been trained in it themselves.

- **The followers**: were trained by members of the discipline as they studied under their guidance or helped them in their research work.

Ben-David and Collins traced the lineage of these scientists and their analysis identifies a starting point in Germany, where an efficient communication network had developed. This was the result of the determination of several scientists struggling to gain recognition of their social status, in a context where the race for prestige was on. The founders of scientific psychology were physiologists who migrated towards speculative philosophy where it seemed easier to obtain a professorship. However, in so doing, these migrating physiologists lost some of the social prestige attached to the initial discipline; at that time, physiology was more prestigious than philosophy. They came up against a role conflict owing to the fact that they had moved from a prestigious role to one that was less prestigious. Such role conflicts can be resolved either by accepting the loss of status and identification with the former group or by transforming the host role, which is what the scientists in question did. They introduced recognised physiological methods to philosophy with the deliberate aim of creating a new role, via the hybridisation of pre-existing roles. In this way, a new professional role emerged, which was the cause, and not the consequence, of growth in scientific production.

Leaving behind a prestigious role in order to create a new role based on a related discipline can only work if several individuals join in the movement and if young researchers, who have not yet chosen their field, are attracted by this new role. According to Ben-David and Collins, simply hybridising ideas (which is what happened in France and in Great Britain), is not enough to create a sustainable movement; it must be associated with the hybridisation of roles. In France, where the demarcation between disciplines was less clear and appraisal was more diffuse, competition to obtain prestigious positions was not the same. The need to create a new professional role was therefore less strong. Local arrangements were enough to pursue new research; the existing positions allowed those holding them to turn to their preferred type of research; the sociologist Emile Durkheim held a chair in pedagogy while the anthropologist Lucien Lévy-Bruhl was a professor of philosophy. The Collège de France offered few career possibilities, was not involved in the training of disciples and created chairs to match individuals rather than disciplines. In this context, the institution made individual innovation possible, but did not encourage the emergence of new professional roles.

Matching institutional resources and conditions

Simply inventing a speciality does not guarantee that it will be sustainable. Its durability depends on how attractive it is, especially in the eyes of young researchers. New specialities rarely incite confirmed scientists to convert (Stehr and Larsen, 1972). The possibility of teaching and supervising the work of young colleagues proves to be a decisive factor. Firmly establishing a discipline depends on the position occupied in an institution and the kind of academic visibility enjoyed. The 'Phage Group' in molecular biology became a speciality when it began to organise research seminars and entered academic structures (Mullins, 1972).

Institutional entrenchment is a survival condition for specialities whose content would otherwise be ignored or taught under the leadership of a related discipline with a different perspective. The establishment of a discipline requires battles to be fought on different fronts with respect to other specialities, amateurs, institutions and society (Box 3.1).

Box 3.1   *The case of geology*

When, in 1820, Charles Lyell (Porter, 1977) started to study the history of the earth, this field of research was connected to theology, biblical exegesis and palaeontology. Those with knowledge on the matter were specialists who controlled libraries and who were seen as an authority when it came to the 'rational history of creation'. Answers to questions about the age of the earth had already been settled and were hardly ever subject to controversy. Lyell, who was an amateur and a newcomer to the field, had to limit his research to the study of rocks and fossils as he travelled and drew up reports addressed to scholarly societies created in order to set up collections.

However, Lyell put forward new hypotheses about the history of the earth, notably suggesting that it was far older than exegetists thought. As he attempted to make his arguments heard, while still having little geological proof, he found it difficult to refute the affirmations based on a rigorous exegesis of the Bible. The specialists, Cambridge University clerics and scholars awaiting prestigious appointments (bishop or professor of ethics or philosophy), countered Lyell with unquestionable arguments. As for the amateurs, who were passionate but individualistic collectors, they showed very little interest in Lyell, who lived off his father's allowance.

He thus approached the enlightened nobility and spent part of his time giving lectures in fashionable circles. This helped him to earn a living but also meant that he wasted time instead of furthering his research on erosion. It also meant that his position was unclear. To please and continue to interest his public, he said the earth was 'younger' than he actually thought it was by several million years. This was so as not to shock the erudite nobles who believed the earth to be only a few thousand years old. To ensure that he had a more regular income for his science, and to be less subject to the effects of fashion – electricity, magnetism and anthropology were also meeting with much success at the time – he applied to the state, basing his arguments on the fact that his science could help locate new coal deposits, map the country and prepare new land for use by man.

To be more convincing, Lyell had to put together more observations and data. He attempted to convince colleagues, organise collections, create new reviews, define new working norms, push aside amateurs and organise a circle of specialists. As the circle was too restricted, he tried to promote his theories among a larger public by publishing a basic work: *Principles of Geology* (published in three volumes in 1830–33, by the publisher John Murray, London). Purged of overly technical details and using metaphors to help with understanding, the book was likely to be seen as the work of an amateur, when what Lyell was really aiming to do was to professionalize the job of geologist. Outlining a new conception of the history of the earth, Lyell also risked being too far removed from admitted theories. In an effort to institute his new discipline, he waged war simultaneously on all fronts (Latour, 1987). (i) He attempted to keep amateurs at bay while at the same time using them as a work force and providing them with a disciplined framework. (ii) He strove to satisfy polite society while trying not to waste too much time debating their opinions. (iii) He worked on convincing the state that geology was indispensable, without falling into the trap of making any false promises. (iv) He applied to the state to institute a profession of paid geologists but strove to ensure that there was no governmental mismanagement of this profession. (v) He fought against theories and the monopoly of university professors while attempting to introduce a means for his own theories to be taught. At the time, geology had not yet acquired a stable and interested public, or any regular resources. There were no colleagues to examine each other's arguments and put them to the test. There were no amateurs disciplined by a theoretical and methodological framework involving the systematic and/or targeted collection of samples. There were not enough data or accepted theoretical models. No expertise had been built up and there were no laboratories.

The work needed to set up a discipline involves understanding the researcher's institutional environment, but it also means taking a more detailed look at activities, working practices and the content of the ideas put forward.

The discipline as an institutional stake
New disciplines are regularly created but very rarely disappear. Those prophesising that chemistry would be absorbed by physics or biology by chemistry were wrong. In 1964, the *Grand Robert* dictionary listed 150 major disciplines. Over 40 years later, these disciplines still exist while new specialities can be seen to emerge at the interfaces of existing disciplines.

Research managers also tend to group together certain disciplines (spatial sciences, life sciences, engineering sciences, human and social sciences and so on). This offsets the tendency to subdivide and break up disciplines into specialities. Disciplinary divisions and the titles awarded to disciplines are sometimes reviewed. In other words, disciplines are reclassified. Thus, in 1971, when Grenoble University was split into three separate establishments, psychology was classed with the University of Social Sciences while geography was put with the University of Sciences (that is, natural sciences). Ten years earlier, psychology had undergone the opposite movement in the CNRS (National Centre for Scientific

Research), as it was moved from human sciences to join the life sciences department. Sometimes no appropriate point of attachment can be found for a speciality and its visibility is reduced, which explains why classification is an institutional stake.

Indeed, the stakes are high when disciplines are brought together. One of the most notable challenges is that of recognition. The arguments used to justify moving psychology from human to life sciences were based on obviousness, modernity and the scientific nature of this discipline. A more in-depth analysis reveals that the protagonists were also motivated by other reasons:

- **Metaphysical**: the human being studied in psychology was decreasingly thought of in humanist and philosophical terms, based on introspection. The human being was no longer studied from a medical or clinical point of view but as a living being, like any other living being to which life science methods could be applied.

- **Epistemological**: the methods used in psychology were closer to the experimental sciences and the production of phenomena in controlled conditions.

- **Sociological and economic**: recovering a certain amount of legitimacy thanks to this recognition as scientists, researchers in psychology could obtain new resources in order to create experimental psychology and cognitive psychology posts. The poor relation in all of this is clinical psychology, which attracted all the students but was regarded with an increasing amount of condescension.

Disciplines are also co-built and co-evolve. The history of 'economic and social sciences' in French secondary education can be compared to a political combat involving pressure groups (parents' associations, teachers' unions, professional teachers' associations belonging to other disciplines), and various authorities (bodies of inspectors, education ministry, conference of university presidents). The arguments put forward focused on overloaded training programmes, the scientific nature of the disciplines, the possible merging with other disciplines such as history and geography, the pedagogical approach (the pupil as a co-author of knowledge), and so on. Making room for a discipline means that the space taken up by other disciplines has to be reduced. Such battles involve legal and technical constructions, the encoding of teaching content, the definition of forms of assessment and the setting up of procedures and criteria to recruit teachers. The knowledge taught bestows a rating on the institution and enables the discipline to be socially recognised.

This is the kind of skirmish experienced by Durkheim and his disciples when striving to institute sociology. Their battles were waged through professional associations and reviews. They wanted to be recognised in university and research institutes, be able to set up autonomous training programmes and create posts. In 2004, the list of university teaching vacancies in France included seven in anthropology, 47 in sociology, 99 in economics and 158 in management. And so the battles continue throughout the recruitment process, with the question being who should actually be recognised as belonging to the discipline. Much depends on how open or closed in on itself the discipline is. This type of debate is sometimes about the recognition of a title (engineer, doctor, psychologist and so on), or about the recruitment conditions for a given programme.

Taking cognitive content into account

Considering institutional factors in order to understand the emergence of a discipline leads to a study of the academic structures within which the new speciality is trying to establish itself. These structures are also linked to cognitive content. Cole and Zuckerman (1975), using the example of the sociology of science as an emerging speciality, made a distinction between two types of specialities:

> - **Those that are built up in opposition to established theoretical or methodological positions**. These stir up much resistance from established scientists, who cut off subsidies and access to research or teaching positions (and therefore access to students) as well as to publication. The appearance of new specialities also generates scorn and mockery in conferences and publications.
>
> - **Those that emerge from the study of a new object or use of a new method**. In this case, there is limited resistance, which does not necessarily mean that the speciality raises enthusiasm; it can exist in a general atmosphere of indifference.

The establishment of a speciality is conditioned by the structure of the academic world, but it also depends on whether or not scientists actually perceive a problem or challenge. The collective perception of research fronts facilitates the emergence of new specialities, at the boundaries of existing disciplines (for example, the biological molecules of physicists in the mid-twentieth century). The emergence of a speciality can also be correlated with scientific migrations, the decline of other specialities and the grouping together of institutions or individuals locally.

The subject of the birth of scientific specialities captivated sociologists until it was dethroned by the analysis of controversies and the study of laboratories.

The Structuring of Research Fields

Sociology and history have widely contributed to highlighting the emergence and spread of the disciplinary model of the sciences, as well as to the drawing up of monographs about the birth of specialities. Far from Merton's unified model of the scientific institution, the dominating image is rather one of an archipelago of specialities and scientific pluralism. The structuring of research is itself very different from one field to the next. German psychology at the start of the twentieth century and British social anthropology, for example, form a 'polycentric oligarchy' in which a handful of researchers occupy dominating positions and create rival schools, using their own results assessment methods. The situation is very different in Anglo-Saxon neoclassical economics, which operates like a partitioned bureaucracy with standardised training programmes, relative theoretical cohesion and a focus on analytical work (very little control of empirical phenomena). Alongside these two types of structuring, Whitley (1974) points out others:

- **Fragmented adhocracy**: without any kind of collective direction or overall consistency, coalitions in the field are temporary. Objects are defined outside of the discipline according to the audience. Example: British sociology and literary studies.

- **Professional adhocracy**: coordination of research resources for multiple objects and projects, according to audience and influence. Example: biomedical sciences.

- **Polycentric profession**: common working procedures act as a framework for controversies in the field. Example: experimental physiology.

- **Technologically integrated bureaucracy**: coordination of work using the same set of instruments, methods and concept (nomenclature). Knowledge is both empirical and specific. Example: chemistry.

- **Conceptually integrated bureaucracy**: coordination of work via a unified theoretical framework establishing the hierarchy of specialities. Example: physics.

Each field is characterised by a series of structural factors, brought to the fore via the examination of interdependent (Box 3.2) links between researchers and the degree of uncertainty inherent in the field.

What is a discipline?
The sociological definition of a discipline or a speciality is a problem in itself. The main question is whether it is recognisable through specific normative structures. A discipline can also be identified through exchange structures (scientific societies and reviews) or reproduction structures (training programmes, recruitment and review commissions). However, it is not always possible to superimpose the divisions produced using this type of data. Divisions vary according to country, for example research fields structured and instituted as departments such as 'science studies' and 'women's studies'. In France, the way the National University Council splits up disciplines is different from that of the CNRS National Committee, whose aim is to take a different approach from that of disciplinary reproduction.

Information exchange, cooperation and influence systems create links between researchers who share a specialised language, concepts, methods and common tools. These sociocognitive links help define the contours of specialities. Some are easily identifiable (overlapping of sociological and epistemological division modes); others are heterogeneous and their boundaries uncertain. Perceptions differ as to the exact positioning of boundaries while the terms applied are rarely unequivocal 'disciplinary markers'. Terms can be used simultaneously by

| **Box 3.2** | *Characterisation of organisational structures (Whitley, 1974)* |

Interdependence between researchers arises from their need to have their work recognised and be on a par with the work of their colleagues. It varies in relation to three things: (i) the intellectual autonomy enjoyed by the field with respect to society and its ability to impose norms and concepts; (ii) its dependence in relation to specific resources (for example, access to rare pieces of equipment); and (iii) the diversity of audiences targeted by the field. There are two dimensions to this dependence:

1 **Functional dependence (FD)**: the different types of knowledge produced rely on each other. When there is a high level of FD, teams use the same procedures and link up their results. When it is low, the procedures vary and the results are not cumulative.

2 **Strategic dependence (SD)**: work depends on the definition of collective research priorities and the allocation of resources. When SD is high, teams rival with each other in order to control resources and recognition in the field, including control over the name given to the field. When it is low, teams are less concerned with the hierarchy of objectives across the field.

Four cases arise out of the combination of these two types of dependence (Table 3.1).

Table 3.1   *Strategic dependence and functional dependence*

		Functional dependence	
		Low	*High*
Strategic dependence	Low	Anarchy of the teams pursuing different goals with different methods, without either coordination or division of labour?	Pacific coexistence of specialised teams working with standardised methods and coordinating their work but without imposing any structure to their research field
	High	Fight between schools of thinking for the domination of the discipline. Teams internally coordinated but pursuing different goals and using different methods	Fight between subdisciplines (using the same procedures and coordinating their work) for control and the hierarchy inside the discipline

Uncertainty can be split into two dimensions:

1 **Technical uncertainty**: control of procedures and of the ability to achieve results.

2 **Strategic uncertainty**: importance given to research problems, their relevance and degree of priority.

several disciplines: 'mass' is used in physics (weight and mass, missing mass of the universe), in economics (monetary mass), in chemistry (molecular mass), and in sociology (mass communication). They only take on disciplinary meaning in the context of the other terms with which they are associated. Works having a big influence on posterity are also difficult to classify in terms of a single discipline: Joseph Schumpeter, Marx, Herbert Simon, Aristotle (still very much cited in physics), and so on. Disciplines seem more like fluid socio-epistemological arrangements that are born, grow and die off, with their make-up changing over the course of time (Mulkay et al., 1975; Geison, 1981). Several disciplines sometimes study the same object. As for the methods, these are rarely specific to any one discipline (Box 3.3).

Box 3.3   *The experimental approach*

This is not specific to the natural sciences. Mathematical models are used in engineering sciences, universe sciences and social sciences. Description (textual and graphic) is a key tool in anthropology and is still used in natural sciences (geology, vulcanology, ethology, botany). The building of types is used in a whole range of disciplines when they focus on a new family of phenomena for which no model has yet been created.

There is no single overriding demarcation. Elements are interdisciplinary (for example, the notion of order and chaos), while all the overlapping, conceptual migrations and common methodologies complicate any attempt to define the limits simply. A discipline corresponds less to a given sociocognitive perimeter than to a focusing of the scientific field where theoreticians, encyclopaedists, managers and epistemologists attempt to tuck in the edges and set out identity-related differences. Researchers use the notions of discipline and speciality in a rhetorical manner, as part of their specific battles and strategies: 'we've got to stay within our discipline if we are to continue!'; 'we need to get back to the basics of the discipline'.

A discipline is indexed to collective dynamics that tend to create a system out of a set of epistemological, methodological, language-related and organisational elements. It engages in capitalisation and structuring movements that produce a hard core, a hierarchy, subdivisions and classifications. Objects, concepts, theories, methods, reviews, laboratories, researchers and so on are placed on a scale of magnitude or centrality. Because these collective dynamics are ongoing, the structuring they produce is shifting and variable depending on the authorities involved: a committee in charge of reviewing laboratories, a course proposing a summary of knowledge or a philosopher outlining a rational classification of elements that are central to the discipline, with the risk of oversimplifying the diversity of research theories and programmes. Researchers, sitting on national commissions or editorial boards, witness changes to research objects, issues and movements and observe individuals and laboratories as they emerge and those that are in difficulty. Through peer discussion, these researchers form an opinion of what is happening, appraise the various movements and situate them in relation to the history of the discipline.

Disciplines are composite and subject to logic based on fragmentation and recomposition, both within the discipline and at the place where it borders other disciplines. Instances of disciplinary retotalisation, reorganisation and regulation help to construct common language elements and shared concerns. Disciplines force specialities with centrifugal tendencies to compare and contrast their approaches with other specialities. They help take stock of a situation and foster self-criticism. They sometimes lead to the sanctioning of those who have slipped too far sideways from the norm. They are driven by tensions, linked to the exploration of boundaries and internal differentiations. They are contingent while at the same time they strive to build stability and renew themselves over the course of time. Jean-Michel Berthelot suggests they should be understood as a 'historically anchored linking up of composite elements' (Berthelot, 1996, p. 98). He sees their emergence as the 'progressive constitution of analysis traditions' (p. 100), constantly oscillating between past and present. He bases his analysis on the case of sociology, history, anthropology and economics and defines laws according to which the facts, concepts and methods specific to each discipline are established. This work leads to a 'woolly but focused division' (p. 111).

Interdisciplinary aspects
Disciplines often give the impression of being small autonomous and independent empires, which do not always live in peace with each other. They are thus subject to power struggles:

- **Border wars**: for example, between molecular biology and cellular biology.

- **Wars of conquest and hegemony** where, each in turn, theology, physics, molecular biology, sociology,[1] economics, sociobiology,[2] neurosciences,[3] and so on, believe they can rule over all or part of the fields of knowledge, either by setting up a skilful division of labour, or by making sure they are well placed in the tree of knowledge, or by imposing the *right* scientific method on all the others. The temptation at times is to isolate a metatheory (as a systemic theory) or an interdisciplinary science in a quest for a new unity of all sciences, even including religions too.

- **Despoilment** where the idea is to take over a concept from a related discipline and reformulate it so that it fits with one's own discipline.

- **Attempts at eradication**: for example, in the cognitive sciences, the proclamations by the neurosciences of the forthcoming end of cognitive psychology.

- **Caricature, simplification and instrumentalisation**: researchers apply the problems and practices of their own disciplines in order to perceive other disciplines. They confer on these disciplines a role and an image that do not fit.

- **Economic war** when the authorities in charge of allocating resources (posts, subsidies, equipment and premises), take action to weaken rival disciplines for the benefit of their own discipline. This is how disciplines like botany, zoology and physiology came to be in a state of collapse at the end of the 1960s because the scientific managers of the French CNRS redeployed their resources to molecular biology, hence placing other disciplines below the survival threshold.

- **Psychological war and war of attrition** when, without meaning to have a negative effect, some people repeatedly use terms such as 'soft sciences' when referring to the social sciences, which, in turn, retaliate with a less effective 'inhuman sciences'.

Generally speaking, dialogue between disciplines is never very easy. Sociologists, for example, appear to be on the defensive when they are up against economists who display hegemonic and integrative ambitions, who promote the virtues of mathematical formulae and neoclassical axioms, and who are prepared to make only a few marginal concessions with respect to this paradigm and have only a superficial understanding of the diversity of sociological approaches (Menger, 1997). Sociologists adopt a low profile. There are fewer of them, they work on inordinate problems and they belong to a discipline that has been heterogeneous since its beginnings. Moreover, this heterogeneous quality has made them wary of universalising analyses.

Furthermore, interdisciplinary approaches are often denigrated. Taking scientific risks at the boundaries of a discipline is acceptable as long as the risk remains rooted within that discipline. Now, the history of the sciences and contemporary observations show that there are many exchanges and much circulation of people, concepts and methods. Researchers have research interests in several specialities (Blume and Sinclair, 1974). Four types of justification for interdisciplinary work can be found in literature:

- **Scientific creativity**: the results of this are all the more fruitful when unexpected parallels occur. Many discoveries stem from the fact that researchers work outside of their speciality: Lavoisier handled explosives but was also a tax farmer (accountant and tax collector); Pasteur was a chemist; Einstein was an engineer working in a technological invention office. The greatest scholars were heroes, brave enough to question the established knowledge passed on via routine education. Theoretical innovation seems to occur at the interstices and not at the core of introspective disciplines. Paradoxically, in places with the highest density of researchers, there is a smaller probability of innovation per capita. The creation of new disciplines is linked to the impossibility of following a career in established and prestigious disciplines. Morever, there is a greater transfer of ideas when researchers work at their discipline boundaries and cross over, if only slightly, into other disciplines (Granovetter, 1973).

- **Conquest**: knowledge is a constantly expanding empire. Progress occurs at the outskirts of this empire. It is a question of clearing new land, discovering new continents and exploring the frontiers of knowledge. These frontiers are similar to the boundaries of disciplines where researchers are encouraged to work.

- **Relevance of scientific work in relation to the object or issue at stake**: when addressing certain problems, fragmented scientific approaches would appear to bear little fruit. In a socioeconomic context where the concern is to use research for the benefit of society, voices can be heard campaigning in favour of interdisciplinary approaches in order to step over the partitions between sciences.

- **Concrete research work**: when the focus of research is an instrument, an object or a land, it is sometimes necessary to mobilise the resources of several disciplines. In anthropology, for example, researchers doing fieldwork have to get around the problem of the language barrier; they have to decode the way the people being studied talk about the world and classify its constituent parts; they have to understand the codes of perception and translate these into their observer's coding system. When Da Matta (1992) tried to make a note of a few indigenous words (*gaioes* Indians in Brazil), he had to solve problems relating to phonetics. Later, as he was working on a population census and attempting to locate a village on a map, he learnt the basics of geography, topography and scale drawing, and then discovered botanical, topographical and zoological classifications, religion, law, psychology, politics and even medicine. Different disciplinary resources are needed for better understanding in fieldwork.

Several interdisciplinary models can be pinpointed following an examination of the concrete working practices of researchers. These highlight the extensive scope of heterogeneity:

- **The complementarity model**: linking up of complementary skills to form a joint approach to an issue. Rather than just simply juxtaposing disciplinary contributions, the different points at which knowledge is linked up are explored with the aim of making a joint achievement (analysis, problem-solving, use of an experimental instrument). One of the disciplines sometimes plays a leading role, with the risk of instrumenting the others. Disciplinary cleavages are reproduced without displacing the researcher's identity. Sometimes, the protagonists become aware of the limits of their discipline and hence query the divisions and feel the need to revise their conceptual approaches.

- **The circulation model**: researchers belonging to one discipline explore others in order to borrow their concepts, methods, questions or problems to be solved. Lavoisier thus imported tools and methods from experimental physics for the purposes of chemistry; the *École des Annales* (French school of annals), did the same for history, by opening up to economics, sociology and then anthropology. The receiving discipline is itself sometimes structured around several specialities according to the imports made. Pre-history provides an excellent example of this as it draws its resources from anatomy, technology, ecology, genetics, ethology, psychology, sociology, anthropology, chemistry and physics (dating techniques), climatology, botany and zoology.

- **The fusion model**: grouping together of researchers working on the same object, attenuating the distinctions between the initial disciplines. Ecology is an example of a speciality born from the merging of several types of knowledge around an object and concepts such as niche and ecosystem. It brings together knowledge from botany, zoology, pedology and orography. The new speciality reconfigures the identity of researchers around a new reference. In other cases, this reconfiguration only leads to a vague assembly that fails to be recognised from an institutional point of view.

- **The confrontation model**: where existing disciplines enter into debate. The, at times, cutting interactions have a backlash effect on the disciplines: repatriation of joint productions and shifts during confrontations.

Box 3.4   *The case of economic sociology*

At the end of the twentieth century, some research claimed to draw inspiration from economic sociology as a hybrid speciality. This was the result of several traditions: the economic sociology of economists, German sociology, Durkheim's French sociology and the Chicago school of sociology. Historically, this coming together did not occur until after sociology and political economics had been established, with difficulty, as separate fields, in around 1820–60. Sociology was the result of a 'divorce'. This was experienced as painful because each discipline had to specialise in a field that only had limited explanatory scope. Sociology was marginalised in the United States between 1930 and 1970. Economics imposed itself as the social science model. A hierarchy sprang up between the two disciplines, linked to the sophisticated development of mathematical tools. The scientific field became more rigid. Authors thus suggested setting up a scientific field at the interface of sociology and economics. A specific research programme was developed, whose ambition oscillated between working around economic theory or right at its centre.

Regimes of Knowledge Production

The dividing up into professional groups, disciplines or fields (fundamental/applied research) cannot fully explain what is happening today. Authors like Krige and Pestre (1997) suggest that reasoning should be based on the dynamics of research spaces. These are termed 'scientific and technological research regimes', and take into account the directions of research and the markets where dissemination of research products takes place. The authors distinguish between four types:

- **The disciplinary regime**: experts in scientific politics and social scientists (Gibbons et al., 1994) spontaneously reason within this framework. Without realising it, they reduce all research activity to this regime and believe that the sciences are undergoing a major transformation. In this regime, research programmes are established according to the criteria of the disciplinary group, based on the quality of the theory, experimental accuracy, concordance between theory and experimentation and the predicted value of the concepts. The results are disseminated within the scientific community via its own reviews and conferences. The main aim of contact with external organisations is to obtain resources in order to pursue the research activity; such contact is therefore steered by the internal needs of the disciplinary group. Disciplinary communities are structured around stable and easily identifiable organisations such as university programmes, scientific reviews and laboratories. They generate a considerable amount of written work, providing access to their results and facilitating their analysis. The disciplinary regime highlights the inherent value of knowledge and underlines a natural difference between science and engineering.

- **The transitory regime**: in this regime researchers oscillate between two criteria for selecting research themes and two markets for disseminating results: fundamental research (unversity referral) and the socioeconomic world. They move between one and the other depending on their needs. Lately, their preferences, in terms of research content, public and reputation, have focused on long-term, disinterested research and exchange with their peers. However, some of the research activity and part of the professional path are defined at the outskirts of the discipline's institutions, either in relation to other disciplines or to engineering or action (managerial, political, medical and so on). Researchers move around inside these other fields depending on their needs with respect to techniques, data, concepts or alliances. Depending on the case, they make their choices according to the criteria of their initial discipline or to those of their allied discipline. Similarly, their work is disseminated either in disciplinary academic reviews or socioeconomic milieux. The transition between these referral agents can take on at least three forms: (i) *a return trip*, limited over time: the identity remains rooted in the initial discipline, in spite of the excursions; (ii) *a lasting, but circumscribed transition* structured according to the initial discipline (for example, from physics to engineering); and (iii) *the creation of new specialities* attached to the initial discipline: physical chemistry, bio-physics, astro-physics, geo-physics and molecular electronics. To understand the scientific dynamics at work in the transitory regime, study should not be limited to within the space of the discipline – which continues to be an important referral agent – but should focus too on the interfaces and displacements effected.

- **The utilitarian regime**: researchers take on problems arising from economic and social demand (engineering sciences, medical sciences and management sciences). In this regime, the population is highly heterogeneous: technicians, practitioners, advisors and so on. They belong to equally diverse institutions: universities, but also applied research organisations, companies, technical ministries, consultancies, industrial technology centres, professional or citizens' associations. The space in which results are disseminated is created by patents, professional reviews and the media. The researchers strive to meet demands and requests linked to concrete end purposes. The identity of people working inside this regime oscillates between the scientific specialisation and the milieu in which the person is engaged (industry, social actions and so on). The differences between the utilitarian regime and the disciplinary regime became less noticeable towards the end of the twentieth century. This is mainly due to the fact that both regimes often share working methods and instruments. Research programmes are sometimes so similar that, in some sectors, researchers say that they are only a few months ahead of industrialists. Practices in the utilitarian and disciplinary regimes are often similar, although the preferred focus of the researchers belonging to these two categories clearly differs over time.

• **The transverse regime**: researchers in this regime are above all interested in metrology and methodology, as well as in the design and development of generic instruments or protocols. Their work helps to further academic research or contributes to utilitarian regime activities. They do not belong to a specific discipline, but form 'interstitial communities'. They work on instruments such as ultra-centrifugal machines, spectroscopes, lasers or, in the social sciences, on content analysis tools, database management, scientometry and so on. Those involved in quality management, knowledge management, the development of modelling and simulation tools and ethics applied to the sciences belong to this regime. Their instruments are based on general instrumental principles and theories. These are used to form systems (for example, for detection, measurement, control and management), which are open, flexible and versatile. The dissemination of their work (instruments and literature) is via a wide range of university, industrial, technological and administrative milieux and sometimes has universal impact. The members of this regime work between disciplines and between institutions, even if they are necessarily attached to a specific organisation. Their identity is built up through cross-cutting and heterogeneous networks. They infiltrate niches and circulate between various groups. They avoid being specifically associated with one organisation. They make it possible to connect up scientific, technological and institutional fields that are at times isolated from each other. They sometimes help to standardise certain practices across diverse institutions. In this regime, the interstitial community moves through phases where it is open and phases where it is more discrete, the latter corresponding to periods in which instruments are conceived far away from the prying eyes and influence of other groups. This regime is transverse for at least three reasons: (i) the research focuses have multiple origins; (ii) dissemination spaces are diverse; and (iii) the community is closely related to the other three regimes. It thus reduces cognitive and methodological fragmentation.

The Case of Engineering Sciences (ES)

Instituted as disciplines, the sciences were mobilised and translated into concrete theories at the start of the nineteenth century. In France, these theories were applied by *École polytechnique* graduates and engineers belonging to the major state professions (military, public works, mines and industries), or civil engineers trained to work in industry. The question is whether these engineering sciences simply applied existing sciences or whether they were themselves a new science, producing its own knowledge.

Engineers, such as William Rankine in the middle of the nineteenth century, defined a place for engineering sciences (applied mechanics, the study of heat transfer and so on). They set these sciences somewhere between theory and practice (Box 3.5), such that they threatened neither scientists nor engineering design offices. Linking scientific theories with practical applications, the engineering sciences developed around problems encountered by engineers (for example, use of the dynamometer for testing water turbines or problems with water injectors in steam engines, which was the starting point for the development of thermodynamic engineering). Similarly, electrical engineers transformed James Maxwell's electromagnetic theory in order to develop their own theory for induction engines. Engineers also developed methods (parameter variations, dimensionless parameters, methods of approximation and optimisation, quantitative estimations of load distributions and so on), applied mathematics, phenomenological models and theories (for example, the theory of beams, modelling of transistors and electrical motors, modelling of turbulent flows, combustion kinematics and thermo-chemistry).

Box 3.5   *The debate about the demarcation of engineering sciences*

Philosophers have attempted to discover whether engineering sciences constitute specific disciplines, in terms of theoretical corpora, which are apart from other sciences. They have looked at the differences and similarities in the cognitive structure of theories. They have identified their internal conceptual dynamics and structure. They have shown that the strategy adopted by engineers consists in simplifying design problems in order to make the inevitable complexity resulting from the application of scientific theories more manageable. The debate about the demarcation of engineering sciences is also connected to the building of professional identities. In the history of chemical engineering, for example, production chemists have striven to separate their knowledge and profession from analytical chemistry.

Dealing with concrete machines, the engineering sciences have introduced flaws in the edifices of science and created their own fields of research. Their work leads to concepts and supplies academic research with some interesting problems. There would therefore appear to be two types of engineering sciences:

- **The lesser natural and mathematical sciences** that introduce imperfections in scientific models. Fluid mechanics is a kind of hydraulic engineering to which the problem of viscosity has been added.

- **The separate sciences** whose aim is to study and understand the properties and behaviour of human-built artefacts. The study of combustion in heat engines (thermodynamics and chemical kinetics) is a perfect example. The phenomena studied often do not exist in the natural environment.

The legitimacy of 'engineering sciences' *is today a highly controversial subject*. The debate focuses mainly on the official support and recognition that public authorities should award these sciences. Researchers in these fields, with the help of industrialists, push for the creation of specific scientific commissions or departments. This can be seen when there is a political will to line up fundamental research with the economic interests of a country, as was the case with the CNRS in 1975–77. The dynamics involved in the drive for autonomy led to the setting up of an engineering sciences department. But the question still remains: do engineering sciences apply scientific knowledge (for which economic actors and applied research organisations should suffice), or, on the contrary, do they deal with new fundamental questions stemming from technological artefacts (combustion engine, turbine, assembly line, new materials, processes and so on), which do not appear to be immediately related to technological developments and hence justify more academic research? The protagonists of engineering sciences work hard to highlight what is specific to them, notably by drawing up their own disciplinary classifications.[4] They also adopt the disciplinary regime mode: making new disciplines autonomous in relation to industrial issues, publishing in international reviews applying academic quality standards and generally raising the level of abstraction and theorisation.

Conclusion: Change of Balance between Regimes?

Reasoning in terms of scientific and technological regimes makes it possible to overstep the classical oppositions between fundamental and applied research, but also the classical analyses in terms of professional and disciplinary roles and groups. This new conceptualisation underlines the many-sided nature of research and highlights the importance of dissemination spaces and choice of research focus in the way these regimes operate. Historically, the four regimes work together; they are partially interdependent and engaged in a mutual enrichment process. However, within the university system, the disciplinary regime has dominated for two centuries, in spite of substantial transformations. Although the relations between these different regimes have strengthened, the first still carries much weight (Gingras and Godin, 2000). In spite of politically determined attempts to transform the research system over the last 30 years, and the organisational changes that have actually taken place, just how far has the overall balance shifted? Just how far is it likely or able to shift further? These research questions for the sociology of the sciences are also at the heart of the great debates about the future of research.

Notes

1 Comte saw sociology as the chosen science for enlightening the world and indicating the way forward. In the 1970s, relativist sociology of the sciences in Great Britain reduced

the claims of physics to simple social constructions that could only be explained through sociological analysis.

2 Sociobiology attempted to explain behaviour, society-related facts, culture, politics and morals using strictly biological bases.

3 With the concept of 'neuronal man', neurophysiology claimed that it had founded and unified psychology, psychoanalysis and aesthetics.

4 In 1955, a committee of the American Society for Engineering Education (ASEE) distinguished six sciences: solid mechanics; fluid mechanics; thermodynamics; inertia, heat and mass transfer mechanics; electrical theory (fields, circuits and electronics); and study of materials.

Recommended Reading

References in other chapters: Storer (1966) in Chapter 2.

Ben-David, J. and Collins, R. (1966), 'Social factors in the origins of a new science: the case of psychology', *American Sociological Review*, **31**, 451–6.

Berthelot, J.M. (1996), *Les Vertus de l'incertitude. Le travail de l'analyse dans les sciences sociales*, Paris: Presses Universitaires de France.

Krige, J. and Pestre, D. (eds) (1997), *Science in the Twentieth Century*, Amsterdam: Harwood Academic Publishers.

Whitley, R. (1974), *The Intellectual and Social Organization of the Sciences*, Oxford: Oxford University Press.

Znaniecki, F. (1965), *The Social Role of the Man of Knowledge*, New York: Octagon Books.

References

References in other chapters: Ashmore et al. (1989) in the introduction; Latour (1987) in Chapter 8.

Alberts B., White, R.M. and Shine, K. (1994), 'Scientific conduct', *Proc. Natl. Acad. Sci. USA*, 91, 3479–80.

Barber, B. (1952), *Science and the Social Order*, Glencoe, IL: Free Press.

Ben-David, J. and Katz, S. (1982), 'Scientific societies, professional organizations and the creation and dissemination of practical knowledge', *Proceedings of the Global Seminar on The Role of Scientific and Engineering Societies in Development, New Delhi, December 1–5, 1980*, New Delhi: The Indian National Science Academy, pp. 153–61.

Blume, S. and Sinclair, R. (1974), 'Aspects of the structure of a scientific discipline', in R. Whitley (ed.), *Social Processes of Scientific Development*, London and Boston, MA: Routledge & Kegan Paul, pp. 224–41.

Cole, J. and Zuckerman, H. (1975), 'The emergence of a scientific speciality: the self exemplifying case of the sociology of science', in L.A. Coser (ed.), *The Idea of Social Structure*, New York: Harcourt Brace Jovanovich, pp. 139–74.

Da Matta, R. (1992), 'L'interdisciplinarité dans une perspective anthropologique:

quelques réflexions de travail', in E. Portella (ed.), *Entre savoirs. L'interdisciplin-arité en actes: enjeux, obstacles et résultats*, Toulouse: Eres, pp. 57–76.

Eisenstein, E. (1979), *The Printing Press as an Agent of Change: Communications and Cultural Transformations in Early Modern Europe*, Cambridge: Cambridge University Press.

Geison, G. (1981), 'Scientific change, emerging specialties and research schools', *History of Science*, **19**, 20–38.

Gibbons, M., Limoges, C., Nowotny, H., Schwartzman, S., Scott, P. and Trow, M. (1994), *The New Production of Knowledge. The Dynamics of Science and Research in Contemporary Societies*, London: Sage.

Gingras, Y. and Godin, B. (2000), 'The place of universities in the system of knowledge production', *Research Policy*, **29** (2), 273–8.

Granovetter, M. (1973), 'The strength of weak ties', *American Journal of Sociology*, **78**, 1360–80.

Menger, P.M. (1997), 'Sociologie et économie', *Revue Française de Sociologie,* **38**, 421–7.

Mulkay, M., Gilbert, N. and Woolgar, S. (1975), 'Problem areas and research networks in science', *Sociology*, **9**, 187–203.

Mullins, N. (1972), 'The development of a scientific speciality: the Phage Group and the origins of molecular biology', *Minerva*, **10**, 51–82.

NAS et al. (1992), *Responsible Science, Volume I: Ensuring the Integrity of the Research Process*, National Academy of Sciences, National Academy of Engineering, Institute of Medicine, Washington, DC: The National Academies Press.

Porter, R. (1977), *The Making of Geology: Earth Science in Britain, 1669–1815*, Cambridge, Cambridge University Press.

Stehr, N. and Larsen, L. (1972), 'The rise and decline of areas of specialization', *American Sociologist*, **7** (7), 5–6.

Zuckerman, H. and Merton, R. (1972), 'Age, aging and age structure in science', in M. Riley, M. Johnson and A. Foner (eds), *A Sociology of Age Stratification*, New York: Russell, pp. 292–356.

4 The sciences as an organisation

There is a surprising paradox in this idea. Sociologists of the sciences have focused much on the institution and norms governing behaviour in science. And yet these are not concerns prevalent in the day-to-day discussions of scientists. However, there is a constant concern about organisation, felt by both scientists and politicians. Scientists wonder whether it is better to organise their activity according to themes, projects, teams or fields of expertise. They talk about how to manage instruments, share technicians among different projects and ensure internal scientific coordination. They negotiate how work should be divided up between scientists, research engineers and technicians, and especially how autonomous or involved each should be. They discuss operational issues in systematic general assemblies or by setting up a laboratory council. They come to agreements about the operational rules for financial management, equipment purchasing, information dissemination and the co-opting of new members. In research bodies, these organisational issues are just as important. They concern the optimal size of laboratories, the incentive mechanisms, the distribution modes (according to discipline, object, and so on), transverse aspects and procedures for allocating resources and carrying out appraisals. From a political viewpoint, the preoccupation with how to organise research dominates discussions with questions focusing on the right organisational forms: support for teams of excellence or programming and contracting around thematic objectives; creation of large institutes or small team networking; and integration of equipment in research teams or dissociation between technological platforms and laboratories in charge of conceptual work.

Organisational questions keep research actors busy at all levels. They struggle to defend certain options, convinced that they are important in terms of scientific dynamics and innovation. The organisational structure affects scientific and technological achievements. Laboratory studies thus take into account the organisational variables linked to the allocation of resources, communication structures and relations between organisational entities. These factors are dealt with as characteristics of the local scientific culture, but they are rarely studied as a subject in their own right. The question of organisation has attracted less attention from sociologists of the sciences, compared with their interest in the scientific institution, careers, social stratification, content and scientific practices. Conversely, organisational sociology has focused little on the world of science.

Nevertheless, the science world can also be analysed as a conglomerate of diverse organisations. These are social forms that ensure the coordination of individuals or groups striving towards a common goal. They involve organisational work, which leads to a structure of authority, division of work and mechanisms of coordination and steering. Conventionally, organisational sociology aims to report on the following:

- The drawing up of **organisational goals**, their negotiation, definition and imposition on all actors, the interactions between actors striving towards relatively convergent goals and their appropriation of the organisational goals. A university's goals, for example, are subject to negotiations and power struggles between internal groups (students, researchers, teachers and administrative staff) and external groups (supervisory authorities, potential employers, elected representatives and pressure groups). They are defined in legal texts, but the balance between missions relating to training, research and the contribution to societal problems still depends on the interaction between actors. The same applies to governmental research organisations.

- The methods applied to **divide up work**, roles and activities and the mechanisms of differentiation and specialisation (between activities and within a hierarchy). In physics, scientists can be said to be either theoreticians or experimenters, which is not the case in other disciplines. In some laboratories, the tasks of researchers, engineers and technicians are clearly differentiated. In other laboratories, engineers publish just as much as researchers while researchers can quite happily take charge of the instruments they need rather than leaving this task to the engineers.

- The **coordination mechanisms**: the authority structure, the rules and procedures, the formalisation of objectives and tasks, the standardisation of skills, tasks and/or results, and the systems of communication and adjustment. In academic research, the authority structure is less noteworthy than in other sectors of activity; researchers enjoy a considerable amount of autonomy and are encouraged to show that they can be intellectually independent. Technicians rarely just execute orders. The structure of authority varies according to the organisation (university, governmental or industrial research centre, small or large laboratory and so on), as do the procedures (quality control and management, project management, allocation of human resources according to themes or projects), and the formalisation of working methods. In science, coordination calls on relatively standardised competencies (via training programmes and recruitment mechanisms) rather than on methods or results (except when it comes to publication procedures).

- The **borders**: the way these are set up and stabilised, how they are crossed and their relations with the environment. In science, organisational borders are even more difficult to identify given that within the same entities (notably laboratories), individuals belonging to different organisations work together (in France, university researchers may work jointly with members of the National Centre for Scientific Research (CNRS), for example).

The notion of organisation is distinct from that of institution, although all organisations have institutional anchor points. Laboratories would probably not exist without the institution of research as a recognised social activity. Scientific organisations are also institutions in that they train a specific type of citizen, able to promote an ideal of rationality, in universities, in companies and in public affairs – going as far as the utopian idea of placing engineers at the head of companies, setting up economists as politicians and choosing scientists to guide the people.

The conglomerate of scientific organisations is profoundly heterogeneous. It includes universities and research bodies, laboratories, governmental agencies and research councils, foundations, large facilities and technological platforms, small research companies and industrial laboratories, scientific cooperation networks and scientific societies, scientific publishers and editorial teams. To understand how the sciences operate, these different bodies need to be analysed too.

In what follows, we shall focus on several organisations reflecting the dynamics of the sciences and of innovation. We shall identify the variety of mechanisms at work, as well as the questions that have attracted the interest of researchers. We shall not include universities in this analysis as these were discussed in Chapter 1.

Governmental Research

Following the Second World War, governmental research organisations were set up in most Western countries. These were in charge of carrying out public interest research, considered as inappropriate for universities, owing to the link with technological development, and not sufficiently interesting for industry. Their mission was to contribute to the development of products and services for public health (for example, the French National Institute for Health and Medical Research: INSERM), transport safety (for example, the French National Institute for Transport and Safety Research: INRETS), national energy autonomy (for example, the French Atomic Energy Commission: CEA), industrial support (for example, the French National Institute for Research in Computer Science and Control: INRIA), and development (for example, the French Institute of Research for Development: IRD). With a staff of over 20,000 people, the CEA is one of the first applied research bodies in Europe and ranks second in France for basic research.

Over the last two decades, most governmental research bodies across the world have been reorganised and their missions redefined (Laredo and Mustar, 2004; Nowotny et al., 2001): clarification of roles, financing agreements, contracts based on objectives, increased accountability, requirement to manage research activities more efficiently and the introduction of quality management. Some have been privatised while others have been closed. Many now operate on the markets and have to compete with research firms, university laboratories or equivalent bodies in other countries. Competition focuses on financing and economic value enhancement. Several of these bodies have been turned into semi-commercial entities whose survival depends on their ability to propose services to their 'customers' (public services, private enterprises, associations of citizens such as associations of people suffering from certain diseases, and foundations), or to the general public via fund-raising campaigns. The notion of applicability has become a dominating preoccupation as these bodies are increasingly being required to serve society and turn their attention to the economic value of their products.

In the field of electronics, for example, an organisation like the Electronics and Information Technology Laboratory (LETI), which is part of the CEA, is a typical illustration of governmental research supporting economic and industrial activity. The laboratory was created to generate economic activity by allowing young researchers to develop a technological concept and enhance its industrial value. The young researchers were supposed to leave the organisation after working on their project for a year, thus ensuring a transfer of knowledge to the economic sector. Currently, around 1,000 people work in the LETI. Their mission is not so much to produce academic knowledge us to work on operational concepts and the associated know-how for intermediary users, mainly industrialists, developing new products or services.

Because scientific dynamics can depend on the internal organisation of a given body, we shall take a closer look at the CEA. Devoted both to applied research and basic research, CEA staff members are shared between several divisions. The basic research division, for example, employs 2,000 people who work in fields such as elementary particle physics, condensed matter physics, theoretical physics, physical chemistry and universe sciences. Other divisions work on life sciences or technological research. The basic research division is subdivided into departments (of around 500 people), units (100–200 people) and laboratories (20–30 people).

Hierarchy and Allocation of Internal Resources

There is a highly marked hierarchical structure at the CEA, even in basic research where scientists' autonomy is considered important. In the unit studied by Rosental (1991), the budget was shared out between the laboratories in a unilateral and authoritarian manner by the unit head up until the 1970s. To obtain financing, researchers had to bring a 'petition' before the head of their laboratory. This person then decided whether or not to pass it on to the head of the unit, who in turn decided whether or not to grant financing, without having to justify their decision. Owing to budget restrictions in the 1970s, together with claims from the unions, the head of the unit began to involve researchers in the decision-making process. From this time on, budget negotiations have required the laboratories to formulate their needs. They involve informal discussions with the unit head, decision-making meetings and, at times, appeal procedures. The laboratories' demands reflect their scientific objectives and their needs, but also what they anticipate as being feasible. The laboratory members have internalised the rules and methods of organisation in order to formulate their scientific objectives. They talk about 'self-censure'. The problem for the researchers is to set up projects that can be justified before peers and management. In order to justify a smaller budget, the head of the unit sometimes suggests that researchers 'drag themselves away' from their experimental work in order to devote some time to publication. Inciting researchers to publish is one way of keeping within the budget. In the meetings where decisions are taken, teams can be seen to confront each other as they defend their interests. Because decisions have to be made about scientific projects belonging to different fields, the challenge is to convince the decision makers that a given programme takes scientific priority over others. This change in operating rules has also affected researchers' possibilities in terms of actions. Thus, the fact that substantial investments (greater than €30,000, for example) now require the approval of the department head rather than the unit head means that the latter has less influence on research directions.

Internal organisation also involves the appointment, appraisal and promotion of people. Here, appraisals are structured around a change in position and decisions about promotions. Promotions, for example, are decided during annual meetings chaired by the unit head and where the examining board is composed of the laboratory heads. The laboratories decide on promotions together, unlike in the CNRS where appraisals are performed by a national body. A researcher's career is built very differently depending on which research body they belong to. The rules specific to each organisation can wield considerable influence depending on whether or not they promote seniority, mobility, publication, contribution to priority research programmes or autonomy. In the CEA research unit mentioned above, publication is the most important criterion for researchers' career advancement. The researchers in this unit can be heard to say that they are 'reaping esteem'. In other units, the difference between scientists, engineers and technicians is not as clear-cut, especially when the priority is patent-filing.

Recruitment and Career Management

In order to understand the dynamics of a research organisation the recruitment mechanisms also need to be examined. In the CEA, employees are rarely laid off and much attention is therefore paid to recruitment. As in many research organisations, researchers are recruited after a long observation period (internship, doctoral and post-doctoral studies) and a socialisation phase. This can lead to the researcher being co-opted by the other researchers already working in the organisation. Furthermore, much importance is given to diplomas. Students who have studied in the French *grandes écoles* are often given preference over those graduating from a university. CEA researchers also lecture students doing research Masters. This is so that they can pick out the best students; those able to imagine 'their own research programme', bear 'the anguish of being an explorer' and call a halt when things get off track (something which management is rarely able to do). The subjects of theses, used to attract young researchers, are defined so as not to be overly specific or basic, to prevent the student from being unemployable on the job market; only one-third of them stay on at the CEA once they have finished their thesis. Those applying for positions at the CEA are attracted by the job security provided, but also by the work which is less stressful than in a company, the social benefits, the laboratory's material resources and, for the technicians, the creative, non-repetitive nature of their work, in a team, without too many overseers.

Understanding the way this research body is organised also means looking at the way new positions are opened and careers managed: adjustments in line with population growth, structural balancing, positions created for applicants who are already known and appreciated, possibility of promotion to superior functions or conversion to a different subject area. Some researchers have set up such substantial scientific and industrial networks that their personal influence within the organisation is sometimes greater than that of the unit heads. Some manage to impose their scientific choices without being in a management position and can even do so when they have retired. The extent of social relations enjoyed by researchers is sometimes more important than the executive power of their bosses.

Cooperation and Preparation of Scientific Choices

Characterising a research body not only consists in describing its structures or the mechanisms for allocating resources (both financial and human) and managing them, it also involves examining the concrete work organisation methods and the cooperation mechanisms. In the case of the basic research unit described above, this involves organising seminars for researchers to talk about their work with one another. As for developing science policy, this is something that 'happens just about everywhere'. It involves a large number of internal actors (who may or may not be members of the hierarchy) as well as external actors (mainly industrialists). Although the hierarchy wields more influence here than in the CNRS or in French universities, management's action within the CEA also consists in attempting to bend the main strategic goals. This entails introducing cautious changes to the balance of resources allocated as well as making several exceptional resources available for incentive purposes. The effects of such action grow smaller the further down the hierarchical ladder it is situated. For the unit head, it is important to have a good understanding of laboratory activities to be able to justify the decisions made. However, the unit head's role is above all to ensure that the slightest changes operated are consistent, explained and shown to be legitimate. This does not necessarily allow heads to impose their decisions. By appealing to 'scientists', who may belong to the hierarchy or not (advisors and experts), researchers have powerful opposition forces at their disposal. They enjoy a considerable amount of autonomy when it comes to defining research directions. Their management trusts them in this as long as they have gained credibility through publications. The researchers have the power to put forward convincing proposals to their bosses, who are themselves scientists rather than administrators. 'You cannot give orders to researchers', says the unit head. When researchers are in the wrong, they can only be brought back on track using structured arguments. This is a time-consuming business given that the researchers themselves work on formulating convincing arguments and defending their ideas. In this organisational universe, the battle is mainly rhetorical; it is all about convincing. Not everything relies on rhetoric, however. The unit head has the ability to appoint 'brilliant' members of staff to work on a project and hence weigh in their favour, open up new possibilities and strengthen a research field.

The above CEA example shows several organisational mechanisms at work, often linked to the social mechanisms of scientific communities. The researchers set up networks that cut through the organisational divisions. They report to communities who define research priorities and encourage their members to use the resources of the bodies in which they work. The influence of these communities is all the greater given that the body gives credit to those who publish and are recognised by their peers. The scientific community is thus able to influence research directions via scientific councils where external researchers sit.

Industrial Research

Not much is known about the way research in companies is organised when in fact it represents a considerable proportion of the research work carried out across the planet. Research, as an activity, has also undergone considerable changes in the industrial context. Over the last few decades, it has been reorganised to become more trade-worthy 'internally'. It is required to prove that it is profitable and invited to become independent. It is subject to multiple partnership agreements between companies and with public research laboratories.

In the following subsections, we shall first see that industrial research has a history too, and that this history has a substantial organisational component. We shall take a look at the role played by research within the company and characterise the know-how developed by research directors in industry.

The Introduction of Science in Companies

The introduction of science in companies is closely linked to the industrial revolution and to the technical changes modifying the carrying out of work. During the second half of the nineteenth century, industrial laboratories contributed to the diversification of firms in European countries. The company Bayer, for example, founded in 1863, devoted its activity to the manufacture of dyes. Feeling threatened by the drop in prices in its sector, it recruited the chemist Carl Duisberg, in 1884, and created a laboratory to perform routine analyses and checks. In 1889, Duisberg redirected the laboratory towards research and launched programmes with a dozen or so other chemists. In 1900, he became a member of the company's management board; there were 144 chemists on the laboratory staff at that time. This injection of scientists into the company led to a change in production direction. With the basic substances of antiseptic products being close to those used in dyes, the company turned to pharmaceutical production.

The introduction of research in industry put two categories of actors in the limelight – engineers and scientists – whose links and activity proved to be determining factors for industrial development. By applying their scientific knowledge, they were able to offer new solutions to old problems (Bowker, 1994). The organisation of industrial research depended on requirements relating to production and collaboration between engineers and researchers. One form of organisation associated the industrial laboratory with production problems, with the engineers acting as spokespersons. Research programmes used the skills of scientists and engineers, but also those available in production workshops. Thus, autonomous research programmes were launched and resulted in a specific product: the patent.

The Place of Research in the Company

Patents became the main product of industrial research; the idea was to 'ensure the continuous production of these patents, or weapons' (Bowker, p. 479). Patents provided companies with a means of controlling their competitors. However, the ultimate aim of the patent moved industrial laboratories away from production problems. This movement gained force with the strengthening of generic research (cutting across several company productions) and the grouping together of laboratories. Research became isolated from production and set itself goals that were relatively independent of the rest of the company. It also moved closer to public research through publication and research partnerships. Companies entrusted public laboratories with part of this research: for example, the French National Institute for Agricultural Research (INRA) for the seed industry and the LETI for microelectronics.

In the 1960s, major industrial groups set up substantial research centres. These were like foreign bodies that were difficult to integrate into the overall production activity. This movement gathered speed as new technologies emerged requiring an even greater cross-cutting approach. In the 1980s, the overlapping action of research policies and industrial policies placed scientific research in the trading game and on the 'socially useful' list (employment, competitiveness, efficient use of public money, and so on). Industrial groups came to occupy a strategic position in public research contracts and became an obligatory point of passage for other industrialists in the sector. The central laboratories were closer to universities than to their sales and production departments. They allowed their researchers to do some 'unofficial' research and publish their results in order to keep them in the company; the fear was that if they were kept tied to economic requirements and technical support needs they would flee towards public research. In companies using traditional technology, there were not many researchers in management; research directors held low positions in the hierarchy while research strategy was not taken into account when the industrial strategy was defined. The major strategic decisions were taken by managers with few technological skills. The separation between managers and researchers was exacerbated by the fact that a career in industrial research was rarely seen to be of value. Researchers were assessed according to their scientific performance and their management skills, but not on the basis of their ability to transform scientific output into industrial performance. Research was said to be indispensable for the future of the company, but industrial researchers had to struggle to make sure that this message was heard. It was not easy to demonstrate the effectiveness of research given that its results were not immediate. It was often necessary to wait 5 to 15 years before the effects could be seen.

The function of research today varies according to the industrial sector, the size of the company, its industrial strategy, the place of research within the organisation (whether or not there are researchers on the management board), and the profile of the research director: recognised scientist, manager and 'guardian of the Sisyphean task', innovator or production engineer. Sometimes, research is likened to patronage, which is tolerated on condition that it does not interfere with other company functions and incurs a limited cost. It is part of the company's fixed expenses. Researchers are seen as a necessary evil: people who are difficult and independent, who are not easily shifted and behave like university lecturers. The company's activities are forced to make financial contributions to research, which tends to define its own goals and act as an independent activity. The research function negotiates and receives an overall budget that it manages in a discrete manner. Some activity directors say that they pay their share of research just to be left in peace. Research directors sit on management boards in order to represent research interests, but often avoid importuning their colleagues about problems in their laboratories. Research is conceived as a reservoir out of which some useful innovations should emerge.

In companies where research involves repetitive operations on known products, or where knowledge is codified and stable and work is programmed, the laboratory is an instrument that serves the company and it is the company that defines the objectives of this instrument. In this kind of company, research is associated with production; it is bought and sold with the division for which it works. It is an instrument of action, managed like any other: it is organised in a bureaucratic manner, based on the rigorous definition of tasks, skills, procedures and annual budget. The director of research in this kind of company is someone who executes orders and does not take part in the strategic coordination of the company.

In other companies, there is a distinct effort to incorporate research into the company strategy. Such companies adopt research evaluation procedures, procedures for 'selling' research projects 'internally' to other company functions, and sponsorship procedures; the idea is to make researchers accountable. Because expectations in relation to research can be very high, researchers put up resistance against requests for technical assistance and fight against what they see as 'exaggerated customer focus'. They complain about being constantly interrupted in their projects in order to satisfy requests for help and about being seen as simple instruments. They reproach company management for their lack of long-term vision and for toying with short-term financial opportunities. Conceived as a strategic resource in an economic war focused on innovation, research is managed in a flexible way when it is part of the company strategy. It is managed around projects and teams that are relatively autonomous and whose ultimate goals are negotiated with the rest of the company. Scientists, engineers, salespeople and financiers are involved in coordinating research. Projects are assessed and then possibly financed. They are sometimes chosen for the scientific visibility of the partners associated with the project and for the image they bestow on the company: collaborating with a prestigious laboratory or working on research with a high level of international renown allows companies to focus on their image in relation to competitors and public financial backers.

Industrial research can be carried out internally, externally or through bilateral or multilateral relations, between companies and with public laboratories, at national or international level. Within the same company, there are sometimes multiple strategies: a central laboratory working on generic technologies or cross-cutting problems can be juxtaposed with laboratories specific to each division and the building of a network of public research partners. These multiple research strategies are sometimes the result of a series of events such as the takeover or sale of a subsidiary rather than the consequence of a well-thought-out strategy.

The status of research in a company also reflects the way the company wages war on its markets: strengthening of existing markets, creation of niches for new products, setting up of barriers preventing entry by competitors (the race for patents), and the imposition of new technical standards. In companies able to readjust product consumption norms (action on the market) and production structures (action on the industrial fabric and on the company), the technical and economic stakes behind research activities are redefined.

Box 4.1   *The job of research director*

This profession requires know-how that covers scientific strategy, understanding of people and understanding of the company (Latour, 1991):

The demographist: this type of research director acts with caution when attracting, keeping, humouring, directing and rewarding researchers. 'It's not a question of investing in research but in researchers' according to one director. Bearing in mind that it takes 10 years to form a team of high-level specialists to work on local industrial issues, they can only be managed on a long-term basis. Demographists attempt to relocate researchers who have become less useful owing to changes in industrial strategy. Another challenge is placing researchers inside the company and creating a network of people able to understand the interests and requirements of research.

The appraiser: this type of director selects projects, measures their progress and checks their interest in relation to the company's activities. Appraisal procedures involve regularly meeting with internal customers, upstream of projects, as well as using formalised indicators (for example, the impact of research work on product cost and sales prices).

The match-maker: this type of director prevents research from being marginalised and associates it with other activities.

Changes to Industrial Research Systems

The economic crisis at the end of the twentieth century and the cutbacks applied to structures put an end to many laboratories and led public laboratories to subcontract out research work. This made coordination a very important issue. Because industrialists and researchers were not governed by the same logic and time scales, they had to set up systems combining relationships based on trust, emotional links, scientific credibility and legal procedures. These resources led to symmetry between laboratories and companies in terms of how work was spread out, information shared, materials exchanged, theses co-managed and people moved around. Two coordination modes could be seen: general set-ups with a legal dimension, such as 'CIFRE'[1] agreements, and local set-ups based on domestic agreements, for example, the inter-knowledge arising from a shared past. The CIFRE was seen as an agreement that made it possible to reconcile different requirements. It offered an acceptable time scale for research to be produced, that is, 3 years – not too long for industrialists and not too short for researchers. It also offered the possibility of working on original questions using academic assessment criteria (articles, theses) and taking on board industrialists' concern for technological transfer. The contract was characterised by interactions with hybrid forms of guarantees relating to general rules, resources resulting from a belonging to the same scientific network and a build-up of interpersonal relations. This same meshing of formal set-ups and relationships based on trust can also be found in the creation of companies as spin-offs from research (Box 4.2).

Box 4.2 *Example of company creation based on the build-up of trust in a researcher*
The creation of a biotechnology company using venture capital was part of the dynamics of a network of actors built up around a given technology: a vector of expression for the production of recombinant proteins for therapeutic use. One researcher and the researcher's team started the network when they discovered a system able to multiply certain human substances. Their work led to the filing of two patents. The researcher had three distinct advantages: an international scientific reputation, the successful creation of a transfer centre and an excellent ability to manage industrial relations and negotiations with politicians. One of the actors, the INRA Industrial Relations and Value Enhancement Division, played a determining role in the setting up of the network, in the person of one of its project managers. This person negotiated the patent filing operations and was familiar with both the patented technology and the researchers behind the invention, with whom he had developed a relationship based on reciprocal trust and esteem. When this person left the INRA to work for a venture capital company, which was the subsidiary of a large bank, he stayed in contact with the INRA and with the researchers. Measuring the economic stakes involved in producing recombinant proteins, they saw in the team's technology potential know-how that would justify the creation of a biotechnology company to develop applications. Their trust in the team's ability to succeed in this project was just as determining as the value of the technology itself. The person then made use of their resources and relied on the trust they had built up with the INRA to convince the actors that the stakes were high enough to set up a company.

Companies, public laboratories and state departments also build programmes and networks around innovative projects, governed by cooperation agreements and a coordination system. This has resulted in an interlocking of organisations where the characteristics of national programming logic join forces with logic based on setting up small independent networks. Two forms of coordination co-habit this interlocked space: hierarchical coordination for the overall architecture of the programme and coordination based on trust within the small communities of industrial and public researchers in charge of the subprogrammes. These programmes rely on preliminary agreements between the actors before they can be set up. These focus on the rules of the game, specifying how past achievements should be pooled, how industrialists should contribute financially and who should own the results.

Other Actors Structuring Research

Research is governed by a whole range of different authorities, often working behind the scenes and forgotten about in analysis work. These authorities define priorities and objectives, allocate financial and human resources, define operating rules and organise action. Besides universities, governmental and industrial research bodies and scientific societies, it is difficult to imagine the profusion and overlapping of organisations involved in research.

Agencies, Research Councils, Foundations and Programmes

At European level alone, for example, a whole series of acronyms and abbreviations comes to mind when thinking of technological R&D: CERN (European Council for Nuclear Research), ESA (European Space Agency) and EUREKA. Among the actors that should be studied by sociologists, the following can be cited:

- **Agencies**: organisations entrusted with missions to support research organisations. The CNRS can thus be considered as an agency entrusted with a mission to support fundamental research. Once in charge of structuring and managing national research, it is now considered by many as a 'resource agency', that is, a body that allocates human and financial resources to research laboratories and instruments. For others, it is a 'labelling agency', which – via its national committee – awards laboratories with a quality label.[2] Depending on the country, these 'agencies' are more or less independent and integrated into state departments. In France, in 2005, several agencies were set up: the National Agency for Research (ANR), the Research and High Education Evaluation Agency (AERES) and the Industrial Innovation Agency (AII).

- **Research councils**: support for research sometimes requires the intervention of research councils (for example, the British National Research Council: NRC) or national funds (for example, the Belgian National Scientific Research Fund). These allocate research resources to researchers or teams according to the excellence of their projects, assessed by peer review committees.

- **Foundations**: these can be private or public and play a supporting role for research in a similar way to research councils or agencies. In 1974, European policy created a European Science Foundation (ESF), following in the wake of its American counterpart the National Science Foundation (NSF). Its members are research organisations and agencies in the member countries. Set up to support basic research, the ESF has only ever received a marginal budget (less than €10 million compared with the €3 billion allocated to the FP).[3] It nevertheless plays a coordinating and steering role, for example for the European synchrotrons, the arctic research programme and the neuroscience programme. It also organises research conferences. The Ford, Rockefeller and Volkswagen foundations are private bodies whose research-supporting role provides considerable incentive in some countries. The Ford Foundation, for example, acted as a trigger in the emergence of the European Space Observatory thanks to the financial means that it provided. Some foundations are of a smaller size, such as the Foundation for Human Progress, which supports some development-related research activities. The law on research, debated in France in 2005–06, allows for the creation of scientific cooperation foundations, legal entities set up under private law for non-profit-making purposes, in order to develop research mentoring.

- **Programmes**: these allow involvement in research with the aim of structuring the dynamics at work using incentives. They are managed by structures such as agencies, foundations and public departments (for example, the European Union General Directorate for Research or the Research Ministry in France). Some programmes are linked to interministerial or intergovernmental agreements.

These research actors affect research dynamics through the resources they allocate and the procedures they implement. They help to build societal compromises that govern research activities. The creation of new agencies and foundations is currently turning the research landscape upside down. Sociologists of the sciences will probably devote themselves to an examination of this situation.

Big Facilities and Technological Platforms

The research landscape is also made up of big facilities (or instruments) or research infrastructures that often have a structuring effect. These are the result of researchers' actions on members of the scientific community as well as policy makers. International organisations (such as UNESCO, the OECD and the European Union) and foundations also play a role as triggers or facilitators of these large structures. It often takes several years to build such structures, and this is often after a dozen years or so spent on the decision making and intergovernmental negotiations. This is the case of the CERN whose particle accelerator is based in Geneva, but it is also the case of the European Space Observatory (ESO) set up in Chile, the Institut Laue Langevin (ILL) and the European Synchrotron Radiation Facility (ESRF) in Grenoble. The life of these research infrastructures is often marked by successive generations of machines. Indeed, the history of scientific cooperation in Europe is peppered with such machines.

These large research infrastructures (Box 4.3) are often associated with organisations designed to manage them and control access to them by researchers applying to use them for experiments. Laredo and Mustar (2004) distinguish two models:

- **International organisations** directly controlled by the associated governments. These have their own staff, who benefit from diplomatic privileges. This is the case of the CERN, which employed 2,000 people in 1999 and brings together a community of 7,000 scientists involved in different experiments.

- **Civil societies and non-profit-making organisations**, which are subject to the management rules of private companies. This is the case of the ILL, whose shareholders are the French CNRS and CEA and the German KfK.

Box 4.3 *Big facilities*

Some big facilities belong to international organisations (the Tokamak depends on the European Commission) or are part of multilateral agreements between nations such as the International Thermonuclear Experimental Reactor (ITER), the Institute for Millimetric Radio Astronomy (IRAM), the European transonic wind tunnel (ETW) and VIRGO (gravitational physics detector). Among these big facilities, the European rocket launcher ARIANE should be taken into account. This instrument is managed by the ESA, whose missions are: to guarantee that Europe has independent access to space, to build a high-level scientific programme and to create the capacity to develop its own satellites. The ESA has successfully carried out several scientific operations in the field of astrophysics with the Giotto, Ulysses and Soho probes.

Clusters in which there is a concentrated amount of resources are another type of research infrastructure. This is the case of the European Molecular Biology Laboratory (EMBL) in Heidelberg or the MINATEC cluster for micro- and nanotechnologies in Grenoble. The latter brings together both engineering schools and research institutions. Such infrastructures sometimes give rise to large databases accessible to a high number of laboratories; the study of their dynamics is particularly interesting to study for sociology of sciences (Keating and Cambrosio, 2003).

The increasing use of instrumentation in science is closely linked to new forms of research work organisation. Collective work, which can be carried out by small teams but also by large organisations employing several hundred researchers, goes hand in hand with the increasing sophistication of instruments. This then leads to an increase in the size of laboratories and to the organisation's formalisation. It also leads to the emergence of technological platforms, and to the grouping together of instruments and teams able to use and maintain the equipment, which is made available to private and public research teams. The emergence of these platforms is strongly supported by various scientific policies set up at the end of the twentieth century. The phenomenon is particularly significant in the fields of biotechnology, microelectronics and nanotechnologies where the potentially useful equipment is at once varied, rare and very costly. It is linked to the adoption of industrial work models and public policies that constitute new research spaces.[4] Although big facilities projects may come to fruition, in spite of pitfalls, re-negotiations and scientific or political turnarounds, this is not the case for all projects. In oceanography, for example, large international ship projects have failed.

It is the job of history and sociology to report on the dynamics of these forms of scientific work organisation and what they produce from a scientific and societal point of view.

The Laboratory

The sociology of the sciences for a long time focused only on individual agents: the implicit model was that of individuals competing with one another. However, another model has come to the fore: that of research as a collective activity. The productivity and the quality of the work of researchers who collaborate are overall greater than researchers working on their own. Over a 15-year period (1980–95), the average number of co-authors went up from 2.5 to 3.5. Teams and laboratories form intermediary organisation levels between researchers and the scientific community. The laboratory is a specific and historically dated form of organisation (see Chapter 1). Designed to support education, it was linked to the ambitions of teachers who above all saw it as a means of attracting the best students. In fact, the laboratory has also allowed a high level of scientific productivity. In the space of half a century, it became a model of organisation. The spread of this model helped to speed up research activity development. The next subsections will focus on how to treat this form of organisation.

The Laboratory as a Production Unit

Putting to one side the institutional breadth of the laboratory, the latter may be described as a productive entity that transforms a series of inputs (instruments, publications, samples and materials, high-level human resources and so on) into a series of outputs (research publications and reports, new instrument prototypes and so on). It is a place where activities and processes unfold, such as the conversion of young students into specialised researchers, the building of a solution in terms of knowledge that is formalised at a client's request, and the transformation of samples from nature into data and theories. Processes are made up of basic operations. The question is, how can we understand the work flow and the way in which the activity is coordinated, supervised and steered? The laboratory also has the ability to define a strategy and reallocate resources to themes or projects. It commits to alliances with other laboratories with whom it competes as much as it cooperates. It diversifies its activities, acts on the economics of internal variety (thematic or instrumental) or, on the contrary, focuses on several key skills.

A 'compass card' can be used to characterise a laboratory's activity (Laredo and Mustar, 2000). It shows the laboratory's output according to five families:

- **Certified knowledge**, in the form of articles published in academic reviews. This knowledge is easily disseminated because it is codified. It is reviewed by peers and hence certified.

- **Incorporated knowledge** (that is, incorporated in individuals through education and supervision of PhD students and so on), which is then materialised (through instruments, materials, physical or numerical models and so on). This type of knowledge is only partially codified.

- **Scientific culture**: laboratories help to cultivate scientific knowledge and information through scientific information and popularisation processes, participation in public debates and the involvement of the media.

- **Innovations**: via cooperation with other actors (industry, hospitals and so on) whose logic is different, laboratories convert knowledge and inventions into innovations (products, processes, services and so on) and into technical standards that notably enable economic activities to be coordinated.

- **Collective assets**: laboratories contribute to the objectives pursued by public authorities in the fields of health, the environment, safety (health, food and defence), and transport by producing knowledge, expertise and advice.

The compass card (Figure 4.1) can be used to sketch the profile of laboratories (diversified or specialised) in relation to the fields most important to them. Laboratories are often active in several fields between which synergies are formed. For example, investing in student training is a means of obtaining the resources needed to produce certified knowledge.

The Laboratory as an Organisational Form

Laboratories differ from an organisational viewpoint in relation to:

- Their **size**: from a handful of researchers to several hundred.

- The **division of labour**: between designers and operatives; researchers, engineers and technicians; theoreticians and experimenters; senior researchers, junior researchers and students; those who publish, those who sign the publications and the others; and between specialities. These divisions are at times implicit and at times marked and formal.

- The **coordination mechanisms**: formalisation of objectives and tasks; standardisation of procedures (protocols, quality management, knowledge management, project management); checking of results (validation of publications); cross-cutting activities (number and type of seminars).

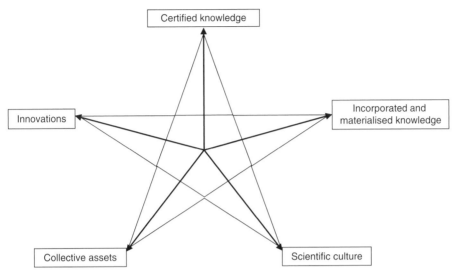

Figure. 4.1   *The compass card*

The question of heterogeneity is complicated by problems of denomination. A laboratory can be a space where experimental research is carried out with a work surface and instruments. But a laboratory is also an organisational entity bringing together researchers, in a lasting manner, within the same management entity, whether there is a work surface or not. In some situations, the shortage of work space is such that researchers do not even have an office; or the laboratory is simply an organisational entity. In this case, it is referred to as a 'joint laboratory' to indicate that it involves pooling resources for a specific topic or problem. Sometimes the notion of laboratory is equivalent to that of 'team' (a temporary grouping together of a handful of researchers to work on a specific theme or project); often it is made up of several teams, and may even correspond to a federation of teams of varying numbers, and include several hundred researchers. There are also 'laboratories without walls', that is, without premises, formed of the association of separate laboratories joining forces for a given period to study a specific problem; in this case the laboratories form a network to coordinate their strategies.

The sociology of the sciences has focused on this diversity of organisational forms and on its causes: intrinsic factors (objects studied, scientific content and instrumental practices), contextual factors (size of the organisation, type of social or industrial demand and so on), or the strategy of the actors who build laboratories.

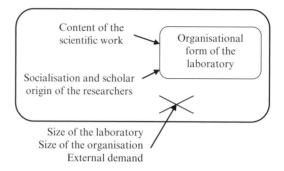

Figure 4.2 *Organisational form determinants*

Factors Determining Laboratory Organisation

Researchers have to meet requirements affecting the way they organise work and their relations according to the type of scientific practice and speciality (Shinn, 1982). A correlation can apparently be drawn between the discipline and the organisational model. Neither the size of the laboratory or organisation, nor the forces behind demand (the market in the case of industrial research) has a significant impact on laboratory morphology. Research units are not affected by the determinism that influences other types of human activity, notably industrial activity. However, the organisational form depends on three factors that are intrinsic to the type of science: (i) the intellectual process, type of investigation and equipment required; (ii) the educational background of the researchers and the socialisation process they have undergone; and (iii) the laboratory's history (immutable once defined by historical contingencies or evolving towards a bureaucratic form) (Figure 4.2).

Shinn demonstrates this using a study of 13 industrial research laboratories. He distinguishes three types of organisation (characterised by their hierarchical structure, form of authority, nature and intensity of communications and staff work schedule), according to the work content and socialisation processes (Table 4.1). There is *a* correlation between the establishment where the researchers have been trained and the type of scientific work organisation in which they find themselves later. On the one hand, they are prepared to enter into specific organisational structures; on the other hand, they tend to reproduce the form of the laboratory in which they were trained.

While Shinn ponders the explanation behind different research organisation forms, other authors (focusing on management, economics and psychology) query the organisational factors that affect research performance (Box 4.4).

The actors are absent from these approaches, inspired by the structural contingency theory. Perhaps they should be put back into the analysis.

Table 4.1 *Types of organisation according to discipline*

	Chemistry	*Information technology*	*Physics*
Organisation model	Mechanical	Organic	Permeable or hybrid formed of traditional (mechanics and hydraulics) and modern physics (electronics)
Form of authority	Centralised (directors make, often unilateral, decisions about report content)	Decentralised and diffuse: collegial decisions taken (management, methodology)	Central and collegial
Hierarchy	Many levels between director and technician with official delegation of power to engineers	Few hierarchical levels (especially symbolic)	Flexible in spite of number of levels
Communication structure	Rigid and formal (official meetings, work memos, etc.); very little non-codified contact (except within the same hierarchical level); little contact with other laboratories	Free and varied (direct contact between technicians and directors; information spread by word of mouth; collaborative projects, including with organisations outside the laboratory)	Formal but with additional support from informal networks
Division of labour	Predetermined work and division of labour according to hierarchical structure (administration and management/creative work and preparation of experiments/performance)	Variable according to projects (no stable definition of tasks; sharing of responsibilities depending on project requirements; semi-autonomous groups)	Stable, but allowing for variances (codified but negotiable practices and scope for initiative)
Career mobility	Limited	Promotion (sometimes accelerated) in relation to performance; open to technicians including research director positions	Extensive in the upper hierarchical levels but limited for technicians
Work content	Hypotheses and deductions based on concrete objects; no theoretical investigation. Repetitive experimenting to test a series of variables	Mainly theoretical: research and formulation of conceptual and mathematical models	Partly hypothetic and deductive and partly conceptual
Instruments	Simple and varied	Computers/paper and pencils. Computer used especially to check validity of models	Advanced and complex, entrusted to research engineers

Table 4.1   (continued)

	Chemistry	Information technology	Physics
Performance of work	Repetitive work for technicians; results passed on to researchers who then draw the conclusions. Directors analyse results, organise procurement, and generally manage. Instructions are top down and results are bottom up	Each project is entrusted to researchers and widely and collectively discussed. Technicians are involved in all creative processes	Engineers take part in experiments and results analysis with the help of technicians (who are also in charge of the more fastidious tasks and equipment maintenance)
Socialisation process	Trained in the most prestigious higher education establishments where they receive lecture-based training. Knowledge and general concepts are consolidated through practical exercises. Students are receptacles for ideas and their assiduity and note-taking are monitored. This type of education makes them prefer a mechanical organisation model. It is thought that laboratories have to be managed by people having graduated from the right schools as they will be able to exercise authority and command respect. Researchers from other educational backgrounds are assumed to be incapable of respecting the hierarchy and have no sense of order	Come from ordinary universities: inductive education, based on practical work followed by discussions in small groups. Intellectual independence is encouraged: possibility of choosing one's own project or even one's own training programme. Students are invited to question the conceptual framework to which their work belongs. Used to informal social and intellectual relations, their preference is for non-hierarchical organisations. Management by engineers from the more prestigious educational establishments proves to be counterproductive in this area	There are many different establishments, some that are even specialised. This variety is reflected in the composition of laboratories as they are obliged to integrate different specialities so that their approach to a problem is all-encompassing. The enormous range of training possibilities (both specialised and general) is seen to be advantageous in the running of a laboratory. Networks are set up inside the lab according to educational background: their semi-autonomous structures make cooperation and competition possible inside the lab

---

Box 4.4   *Factors influencing research productivity*

Individual determinants (age, sex and training) cannot predict productivity very well owing to the collective dimension of research work (Stephan, 1996).

A laboratory's reputation has a positive impact on researchers' productivity (Cole and Cole, 1973).

Laboratory characteristics (organisational mode, financing structure and contracting strategy) do have an influence on researchers' productivity (Joly and Mangematin, 1996). The size of laboratories, for example, has a negative influence on researchers' productivity.

Laboratory composition: the presence of researchers who also teach has a positive influence on the productivity of full-time researchers as they attract PhD students to the laboratory. The mixture of generations within a laboratory also has a positive effect on individual productivity (Stephan and Levin, 2002).

The Right Organisation

Researchers and research organisation managers alike ask themselves what is the right kind of organisation. From their point of view, the organisation is not entirely determined by research topics or scientific practices. The sociology of work and part of organisational sociology no longer focus on contingency factors. They prefer to explore actors' strategies to set up action regulation modes. Already in 1972, Lemaine et al. (1982), drawing inspiration from the works of Michel Crozier, saw research teams as actors and promoters of projects and strategies, notably in terms of the right kind of organisation. These strategies have an impact on the directions followed by laboratories. The type of organisation reflects researchers' response to their research environment, where they can also find inspiring organisational models, and to the objectives pursued.

The strategies implemented by researchers also influence the collective dynamics governing how closely they work with technicians. Their ability to communicate with technicians also has an effect on research strategies, especially when researchers depend on technicians and their instruments. The autonomy and recognition granted to technicians is at the heart of discussions, negotiations and strategies, all of which result in the final organisation. The more independent technicians, who are less motivated by researchers' work – when they feel dominated and exploited by researchers who take over and steal the benefits of their efforts – advocate the professional norms of 'a job well done' and impose these on researchers regardless of the researchers' goals. In this respect, the laboratory comes across as a space that is structured by actors' power struggles. It is also a cosmopolitan space, frequented by researchers, members and visitors on a variable basis.

An analysis of the paths followed by laboratories reveals the peculiar way in which they are structured. The differences observed reflect not only their action-related contexts (expectations and support of supervisory bodies, local landscape and so on), but their internal dynamics (the researchers' collective organisational and scientific project, their preferences in terms of recruitment methods and their definition of academic excellence and so on). The actors not only adapt the organisation to opportunities, but they also focus on specific projects in the long term. Researchers build laboratories and devise cooperation rules, for example, based on tolerable and satisfactory competition. The laboratory is governed by these rules but, at the same time, is a space where the production of rules and projects cannot be separated from contextual constraints and resources (Louvel, 2005).

Scientific Cooperation Networks

Research organisations are not always relatively stable organisations such as universities, agencies, large instruments or laboratories. They may also be scientific cooperation networks resulting from the circumstantial coming together of laboratories and researchers working on a given theme, problem or project. Such networks have existed for a long time, notably within the framework of international scientific organisations. However, the network phenomenon has developed substantially over the last few decades. The creation of networks, sometimes at a local and informal level, has become a voluntary and collective enterprise, promoted via public research programmes. This impulse to 'set up networks' reflects a political wish to organise scientific work. The involvement of the European Union Commission typically reflects the increasing popularity of this new research work organisation mode. As they are much easier to set up than large specialised laboratories, these networks have become instruments of science policy.

Nevertheless, there is no single network model. The differences from one network to the next are considerable. There are five main types of network (Vinck, 1992):[5]

- **'Collection structure' network**: based on the collection and/or processing of data and samples, this network operates with a high number of local actors (researchers, industrialists and doctors) in order to produce data, pictures or samples. These are then disseminated, sorted into groups, compared and kept in a laboratory, a collection, a database or a sample bank. This kind of network makes it possible to work on a wide range of situations (cases, practices, technologies and so on). It is coordinated by managing the dissemination of documents (papers, computer files and so on), and large databases. The network is sometimes subdivided according to region or theme. It is used to monitor a specific phenomenon (notably epidemiological), harmonise practices (for example, medical), or assess techniques. For purposes other than a research project, some of these networks are instituted in the form of a permanent public service.

- **'Forum' network**: social structure in which scientists exchange ideas and results, conceive of projects (bilateral or collective), and agree on codes of conduct. The forum relies on 'conventional' exchanges between teams, that is, seminars and conferences. It structures a scientific community around research issues, objects under investigation or methodologies. Forums come together in places where small specialised communities need to be organised and problems located at the borders between disciplines need to be explored. A forum does not necessarily have to have a laboratory.

- **'Research practice harmonisation' network**: more solid version of the forum. This type of network enables researchers to exchange ideas, but also data, which are made to be comparable and complementary through efforts to harmonise and standardise researchers' practices, language and instruments. Many intermediary objects (Vinck et al., 1993; Vinck, 1999) travel from one team to the next (samples, reference material and protocols) while joint equipment (reference laboratories, shared instruments) ensure that there is greater homogeneity in data processing. The cost of bringing the laboratories in line is such that there is a clear border that gradually separates those inside the network and those outside. In this network, local scientific output is disseminated and easily understood, as well as reviewed and validated by other researchers. Scientific output outside the network remains local, non-reproducible, incomparable and unusable.

- **Network 'radiating out from a centralised facility'**: hard network in which activities radiate out from a centralised piece of equipment (big facility, reference laboratory, test centre and so on), and material exchanges take place. The teams are normally only linked with the centralised facility, which structures the research community via intermediary object dissemination and instrument access management: approach adopted in relation to problems and harmonisation of team practices.

- **'Project structure' network**: made up of teams with varied and complementary skills that intervene at specific moments in relation to the state of progress of the joint project. This network is characterised by strong logistics and the sharing and integration of tasks between teams. This type of network can be found when new treatments and ad hoc equipment are developed.

All of these networks put heterogeneous actors in contact with one another: researchers, industrialists and doctors belonging to different disciplines, sectors of activity and organisations. They do not form assemblies of peers. Nor do they form stable institutions; they constitute flexible forms of coordination. Of varying composition, they are above all transient and linked to the projects that presided over their building. They produce synergy effects between teams and use existing, if dispersed, resources. They make it possible to move, compare and associate local resources and, because of this, they add value to these resources. When they break up, there remain a series of equivalences (same language, standardised instruments and relationships of trust) that can easily be applied to new projects. At European level, they also help with the building of European policies because it is their job to think up ways to work together in spite of their differences (North versus South, large versus small countries). These networks have invented various arrangements to solve the problem of asymmetry between researchers (Vinck, 1996), for example:

- The definition of a hierarchy of statuses within the network: observers weakly involved in the network; active teams; core teams. The endogenisation of this asymmetry allows the network to include a greater number of teams.

- The selection of a limited number of lesser teams by the highest-performing teams and investment in these teams to bring them up to the same level of high performance.

- The international division of work with specialisation of teams according to respective skills. This type of network hardly encourages learning for lesser teams.

- The invention of working methods cancelling out the differences between teams: standardisation of instruments, protocols and qualifications.

- The implementation of methods enabling each team to develop new projects, in line with their local resources and needs, by capitalising on existing learning within the network.

Researchers invent and build networks. Studying them offers a key to understanding how local output is tied up with international dynamics.

Scientific Publishing

Scientific knowledge is not only produced in laboratories, research organisations and cooperation networks. Other organisations play a non-negligible role in knowledge production dynamics. This is the case of scientific publishers. Their role may come across as secondary and peripheral in relation to the activity of laboratories, that is, it seems to be limited to the shaping and dissemination of knowledge. However, under their leadership, knowledge is reviewed, work is critically discussed, and researchers accompanied in their efforts to codify knowledge, select works and, finally, certify knowledge.

Understanding the operational dynamics of scientific publishers requires an analysis of the Editorial committees (who define the editorial priorities of reviews) and the Review committees (who assess the texts submitted for publication). This notably involves exploring how reviews are made and the dynamics behind them, especially the strategies adopted by researchers and editors, the mobilisation of reviewer networks, the type of deliberations and negotiation of editorial policies. Reviews often claim to play a facilitating, if not a structuring role in research communities. Some reviews target specialised communities and consequently have a smaller distribution, while the articles published are less often cited. Other reviews aim to be more generalised, such as *Nature*. These target a wider readership and a high impact factor. Some prefer articles that are illustrated with photos rather than filled with complicated demonstrations. Others focus more on conceptual developments, results that can be widespread or whose implications reach a broader public of researchers rather than overly specific empirical results relating, for example, to overly technological know-how. The editorial policy of reviews influences researchers and editors in terms of what they choose to publish.

The scientific publication phenomenon is impressive and worth the full attention of sociologists. The twenty-first century so far has 160,000 scientific reviews, covering all disciplines. Among these reviews, 24,000 operate with a review committee and publish over two million articles every year. The publishers are either public (university presses, research organisations and scientific societies) or private (including multinationals such as Vivendi and Lagardère). Some publishers manage several hundred or more: roughly 2,000 for Elsevier, 1,000 for Taylor & Francis and Kluwer, and 700 for Blackwell and Springer Verlag. These are followed by Cambridge University Press and Oxford University Press with less than 200 reviews each. These publishers work in partnership with more or less the same number of scientific or professional societies. Many publishers publish only one review. Their strategy affects publication possibilities and has an impact on scientific dynamics in so far as researchers strive to ensure their work attracts the interest of the most prestigious and often-cited reviews, thus increasing their likelihood of being recognised.

Publishing actors also influence availability and access to research results. In the 1990s, major commercial publishers practised inflationary price policies, one of the consequences of which was that many libraries put a stop to their subscriptions. The world of academic research and university libraries responded by drawing up the Public Library of Science petition in January 2001. The petition was signed by several tens of thousands of researchers and demanded that publishers create a worldwide public library offering free access to research results. In 2004, library consortia, whose budgets were drained by their subscriptions to reviews, announced that they would support the deployment of an online server to be used by researchers to deposit their writings hence rendering them accessible to all. The idea of 'open archives', launched by physicists in 1990, was supported by institutions upset by the fact that research financed with public funds was made accessible to researchers only in exchange for costly subscriptions to scientific publishers. Libraries also joined forces to form consortia in order to limit the arbitrariness of the prices practised by publishers.

New electronic services have also emerged: online publication and consultation of references, abstracts and articles; the ability to move from one article to another via references. These services are also in the hands of large publishing houses. The software they use to link up publications with each other allows them to control readers' access to the original publishers. Finally, among the electronic services offered, there are also those offered by databases, notably the Institute for Scientific Information (ISI) in Philadelphia, which lists citations and publishes Current Contents. Because this base provides unequal coverage of worldwide reviews – that is, European publications are under represented – the European Science Foundation has created its own citation index for the social sciences while other countries define their own list of national scientific journals.

Thus, operating on the apparent outskirts of the research world, there are actors, scientific societies, publishers, libraries and research agency managers whose strategies influence scientific dynamics in a way that is yet to be analysed.

Conclusion: The Intertwining of Organisations

The world of the sciences is populated with professional or disciplinary communities and research regimes, but also with highly diverse organisations (agencies, laboratories, organisations and infrastructures, cooperation networks and publishers). The importance of these has only just been touched upon. We have attempted to understand their genesis (how they are determined by actors' strategies, for example) or role (how they are determining in scientific activities).

These organisations have generally been studied separately up until now. However, we have seen just how far they rely on one another, and even the extent to which they are intertwined. Finding one's bearings among them and being able to act, as a scientific actor, is nothing less than challenging. This intertwining and the way researchers find their bearings is in itself a topic worth studying.

The preceding analyses have all been relatively static in that they describe structures and operations. In the next chapter, we shall attempt to enhance our analysis with concepts and approaches that make it possible to report on the dynamics at play.

Notes

1 Agreement binding the state, an industrialist and a public research laboratory and based on the doctoral thesis of a student. The agreement requires the industrial company to employ the student and requires the student to do their research work in the company.
2 See the works by Picard (1990) and Vilkas (1996) for further information about the CNRS.
3 The European Union Framework Programme for Research and Technological Development.
4 See Peerbaye and Mangematin (2005) for further information on the genome research case.
5 Survey on 120 networks (3,500 teams) working on the European Community Commission's 'Medical and Public Health Research (1987–1991)' programme.

Recommended Reading

References appearing in other chapters: Lemaine et al. (1982), Louvel (2005) in Chapter 7.

Joly, P.B. and Mangematin, V. (1996), 'Profile of public laboratories, industrial partnerships and organization of R&D: the dynamics of industrial relationships in a large research organization', *Research Policy*, **25**, 901–22.

Laredo, P, Mustar, P. (2000), 'Laboratory activity profiles: an exploratory approach', *Scientometrics*, **47** (3), 515–39.

Nowotny, H., Scott, P. and Gibbons, M. (2001), *Re-thinking Science: Knowledge Production in an Age of Uncertainty*, Cambridge: Polity Press.

Peerbaye, A. and Mangematin, V. (2005), 'Sharing research facilities: towards a new mode of technology transfer', *Innovation: Management Practice and Policy*, **7** (1), 23–38.

Shinn, T. (1982), 'Scientific disciplines and organizational specificity: the social and cognitive configuration of laboratory activities', in N. Elias, H. Martins and R. Whitley (eds), *Scientific Establishments and Hierarchies*, Dordrecht: Reidel, pp. 239–64.

Stephan, P. (1996), 'The economics of science', *Journal of Economic Literature*, **34** (3), 1199–235.

Vilkas, C. (1996), 'Evaluations scientifiques et décisions collectives: le Comité National de la Recherche Scientifique', *Sociologie du travail*, **3**, 331–48.

References

References appearing in other chapters: Cole and Cole (1973) in Chapter 2; Keating and Cambrosio (2003) and Vinck (1992) in Chapter 7.

Bowker, G. (1994), *Science on the Run: Information Management and Industrial Geophysics at Schlumberger, 1920–1940*, Cambridge, MA: MIT Press.

Laredo, P. and Mustar, P. (2004), 'Public sector research: a growing role in innovation systems', *Minerva*, **42** (1), 11–27.

Latour, B. (1991), 'Le métier de directeur de recherché', in D. Vinck (ed.), *Gestion de la recherche. Nouveau problèmes, nouveaux outils*, Brussels: De Boeck, pp. 499–520.

Picard, J.-F. (1990), *La République des savants. La recherche française et le CNRS*, Paris: Flammarion.

Rosental, C. (1991), *Politique scientifique et organisation politique de la science*, Centre de Sociologie de l'Innovation. Paris: École des Mines.

Stephan, P. and Levin, S. (2002), 'The importance of implicit contracts in collaborative research', in P. Mirowski, and E.-M. Sent (eds), *The New Economics of Science*, Chicago, IL: University of Chicago Press, pp. 412–30.

Vinck, D. (1996), 'The dynamics of scientific intellectuals within the integrative trend in Europe: the case of co-operation networks', in A. Elzinga and

C. Landström (eds), *Internationalism and Science*, London: Taylor Graham, pp. 162–98.

Vinck, D. (1999), 'Les objets intermédiaires dans les réseaux de coopération scientifique. Contribution à la prise en compte des objets dans les dynamiques sociales', *Revue Française de Sociologie*, **11** (2), 385–414.

Vinck, D., Kahana, B., Laredo, P. and Meyer, J.B. (1993), 'A network approach to studying research programmes: mobilising and coordinating public responses to HIV/AIDS, *Technology Analysis and Strategic Management*, **5** (1), 39–54.

5   Social dynamics in the sciences

In Chapter 2, science was explored as a social institution regulated by universal norms. In Chapters 3 and 4, it was seen as being subdivided into different subsets. We shall now turn our attention to the mechanisms behind these social subsets, in other words we shall study the social dynamics of the world of science. Historical sociology reports on the emergence of this world and the deep-reaching mutations underlying it. But there are also social structuring phenomena at work and these too need to be considered.

The first phenomenon is the growth of the scientific population. The number of researchers has risen from 50,000 at the end of the nineteenth century, to one million in the middle of the twentieth century, to roughly 3.5 million at the start of the twenty-first century. And this is only in OECD countries. The spread of this population across the world is undergoing transformations: research has notably become increasingly popular in Asian countries and, albeit in a tardy and unequal manner, it has become more feminised according to discipline. This population is also subject to migratory flows and circulatory movements. Demographers have suggested models to explain generational balances and their transformations. The population of researchers is thus both heterogeneous and evolving. Furthermore, we shall see that stratification phenomena and transactional dynamics are also involved in its structuring.

Social Stratification of the Scientific Space

In his analyses, Merton describes a theoretical moral equality between researchers, but he quickly realises that there are deep-reaching inequalities throughout the scientific institution. The rewards, which are supposed to incite researchers to contribute to the progress of science, are very unequally distributed; the distribution system creates inequalities and 'stratification'[1] within the scientific community (Cole and Cole, 1973). The recognition and visibility of scientists (see Table 5.1) is highly contrasted, with Nobel Prize winners constituting the group with the biggest media coverage (Zuckerman, 1977)

Stratification can also be observed in publications (Box 5.1), a large number of which produced by only a small percentage of researchers.

A Concentrated Recognition Phenomenon

Scientists like to stock up on the most prestigious marks of recognition, which are often premonitory; they announce a researcher's possible running for the

Table 5.1   *Number of scientists, United States, 1972*

Description	Number
Persons self-indicating in the category of 'scientist' during the general census	493,000
Persons recognized as a scientist in the census of the NSF	313,000
Persons recognized in the biographic inventory of *American Men and Women in Science*	184,000
Holders of a PhD	175,000
Members of the Academy of Sciences	950
Nobel Prizes	72

Box 5.1   *Publication stratification laws*

The **Lotka law** (formulated in 1926 by science historian Alfred Lotka), applies to all specialities and stipulates that the number of authors (An) publishing n articles is equal to $1/n^2$ (for authors with a high level of productivity, the phenomenon is even greater ($An = 1/n^3$)). In other words, 1 to 2 per cent of authors are responsible for a quarter of the articles; 10 per cent of authors publish over 10 articles, while 75 per cent publish only one.

According to the **Price law** (Price, 1963), half of the articles published are generated by a number of authors equal to $\sqrt{n}$ (n being the total number of authors in the field).

The **Bradford law** states that there is an equivalent ratio for reviews; most works in a given field are published in a very small number of reviews. According to this argument, the Science Citation Index (SCI) limits itself to looking into 1 out of 15 reviews, with this latter number representing 75 per cent of the articles cited.

Nobel Prize. Roger Guillemin won the Nobel Prize for medicine in 1978 together with Nicholas Wade, but before then he had received the Gairdner International Award in 1974, the Lasker Award in Basic Sciences in 1975, the Dickson Prize in medicine and the Passano Award in medical sciences in 1976, followed by the French National Science Medal in 1977. However, most researchers never receive any of these awards.

Of course, researchers can receive recognition via other channels. For example, they can be co-opted into academic positions. The appointments range from permanent research fellow or lecturer to being asked to take charge of responsibilities in academic institutions, intellectual societies or reviews. Access to employment is considered as a form of recognition. Conversely, having worked in certain institutions leads to other marks of recognition; 49 per cent of Nobel Prize winners carry out their research in only five universities (Berkeley, Chicago, Columbia, Harvard and Rockefeller), and only represent 3 per cent of American university staff numbers (Zuckerman, 1977). Research subsidies, which are another form of recognition, are subject to the same fate; 38 per cent of American research subsidies are allotted to only 10 institutions (Barber and Hirsch, 1962).

Similarly, only a limited number of reviews are considered to be prestigious. Publishing in reviews ensures greater visibility for authors. General interest reviews such as *Nature* and *Science*, as well as non-specialised disciplinary reviews such as *The Lancet* or the *British Medical Journal* covering medical topics, or *The Physical Review* covering physics, are the ones most valued by authors and readers alike. The citations received for articles are an indicator of the recognition awarded to authors by other authors. Only a minority are actually cited. The SCI and scientometry thus measure the impact of articles and their authors. Over a period of 20 years (at the end of the twentieth century), two authors achieved 60,000 citations, around 50 obtained between 25,000 and 50,000 and several hundred reached over 1,000 citations. Following these, the drop in citations is phenomenal: only 0.3 per cent of articles are cited over 100 times; two-thirds of articles (that is, tens of millions) are cited only once at best.

The Mechanisms at Work: A Meritocracy?

What are the phenomena that explain this concentration along with the constitution of a scientific elite? Does this stratification reflect researchers' quality and performance or is it the result of a discriminatory mechanism bent on reproducing an elite, as Pierre Bourdieu shows in relation to the French academic world? For Merton, the unequal sharing of recognition is explained by the reach and excellence of scientific works. The Cole brothers used the (SCI) to carry out a survey on American physicists. The results showed that recognition depends on: (i) the functional importance of science for society; and (ii) the scarcity of individuals able to accomplish a given scientific task. They rejected the stratification theory based on established groups controlling the allocation of resources and recognition according to their own interests. They argued that few researchers receive recognition because the task they are working on is difficult, that only a few researchers would actually be able to fulfil the task, and that, because it is better not to waste resources, only the best should be entrusted with it. This does not mean to say that there is a barrier from the outset: any young researcher may succeed if he/she has the right ability. Science is a meritocracy; recognition has to be deserved and everyone can try their hand at it. If some receive more merit than they actually deserve, competition will soon knock them back down to their true level. There are thus four types of researchers (most belong to the first or fourth types) (Table 5.2).

Table 5.2 *Type of researcher according to production and reception*

		Production	
		−	+
Received citations	−	Silent	Mass producer
	+	Perfectionist	Prolific

According to this theory, the reputation of a university department and the recognition received by individuals are linked to the quality of their work rather than to their productivity. There is a correlation between the number and the prestigiousness of rewards. The quality of work explains the visibility acquired, which in turn explains access to prestigious departments. Reputation and position occupied have a reciprocal influence on each other. The concentration of publishers, the SCI and subscriptions in libraries, backs up this phenomenon.

The Matthew effect: theory of cumulative advantages

The more scientists are recognised, the more recognition they receive. When two researchers co-sign an article, readers tend to talk about only the best-known one and gradually forget the other. In scientific collaboration, the most eminent researcher acquires more prestige than his/her peers, even if this researcher played only a secondary role in the work. Being better known, he/she tends to receive more prestige for the same quality of work. Merton (1968, 1988) refers to this phenomenon as 'the Matthew effect' in reference to the following verse from the bible:

> For whoever has, to him it will be given, and he will have abundance; but whoever has not, even what he has will be taken away.
>
> Matthew 13:12

The experiment that consisted in submitting a dozen or so articles by prestigious authors to reviews, and then re-submitting them under the name of unknown researchers, is enlightening. Only three articles out of the 12 were identified as re-submissions. Eight out of the remaining nine articles were rejected owing to methodological failings.

According to Cole (2004), the Matthew effect also applies to multiple discoveries and 'discovery groups' – discoveries that are linked to one another and published in various articles by different authors. Out of these sets, only the discoveries made by well-known researchers attract attention.

Recognition attracts recognition. It is spontaneously awarded to scientists who are already held in esteem. Conversely, unknown researchers find it difficult to have their contributions recognised. This phenomenon is a dysfunction for individuals, who are penalised at the start of their career, although it does fulfil a latent function from the point of view of the system; it makes it easier to identify the best articles, making them even more visible. Norbert Wiener (1956) commented: 'I was quite aware that I was an out among ins and I would get no shared of recognition that I did not force.'

The phenomenon of cumulative advantages varies according to disciplines and in relation to their internal dynamics. In recently created fields, which continue to grow and are highly evolving, young researchers can quickly earn recognition and dethrone their predecessors. In established fields, on the other hand, where there is little obsolescence in terms of results (such as in the social sciences), the founders benefit from a build-up of recognition.

Merton emphasised the Matthew effect using the work of the Cole brothers. However, the brothers themselves showed that the citations received in the

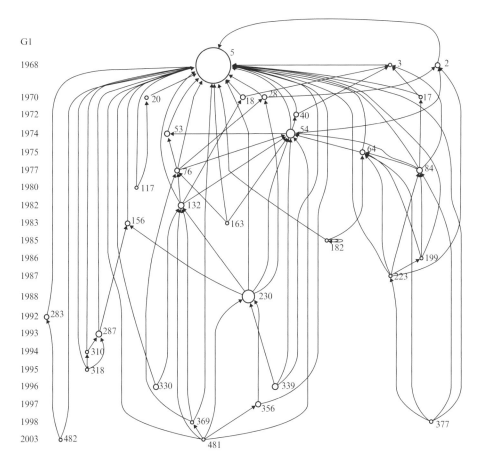

Note: Eugene Garfield, the creator of the *Science Citation Index*, identified over 400 citations, illustrated in this graph of relations.

Source: See http://garfield.library.upenn.edu/histcomp/merton-matthew-effect-I-II/graph/1.html.

Figure 5.1   *The success of the Matthew effect notion*

period following publication (early recognition) did not confirm the Matthew effect (effect of the position acquired on early recognition). Merton did not take these results into account as they contradicted his own theory (Cole, 2004). Along with others, Merton even used Cole's text to back up the Matthew effect hypothesis. Cole explains this denial by sociologists' tendency to look for victims to defend.

Trajectory effect and halo effect

Other effects complete the Matthew effect. The 'trajectory effect' qualifies the advantage that a researcher might draw from the type and prestigiousness of the institutions in which they have studied when applying for a new job (Allison and Long, 1990). In the case of those undertaking post-doctoral studies (in France 30 per cent of those with a PhD do a post-doctorate, 56 per cent in natural sciences and 9 per cent in human and social sciences), the result differs according to the type of institution:

- The effect is **positive** when the post-doctorate corresponds to a job in public research abroad – in biology, this has become more or less compulsory in order to access public research. Moreover, in France, PhD students are more likely to gain access to post-doctoral studies if they do their thesis in a laboratory belonging to the CNRS and benefit from public financing, unlike those who have done their thesis in a university laboratory or in an applied research organisation.

- The effect is **negative** in terms of accessing a teaching post because the specialisation required and the value of research work at post-doctoral level distances the student from teaching networks.

- A post-doctorate has **no effect** in terms of accessing employment and salaries in the private sector.

The 'halo effect' qualifies the advantage of being in a prestigious institution; the researcher's quality of work and productivity are aligned with that of his/her colleagues (Crane, 1965) because they are in a productive scientific environment. The collective resources of research units (that is, scientific networks, skills and equipment) have a greater impact on researchers' productivity than their personal resources.

Constitution and reproduction of an elite: coupled careers
Using these mechanisms to build up recognition, scientists form an elite, attract resources and work in research decision-making bodies. Their productivity at the start of their career allows them to have a say in the allocation of resources and to maintain their advance. This stratification introduces a bias in the competition by putting a store of advantages and rewards into the hands of an elite, hence perpetuating and strengthening it. Fifty per cent of Nobel Prize winners are the apprentices of other Nobel Prize winners. Future Nobel Prize winners learn to use an aesthetic criterion during their socialisation process. This criterion enables them to weigh up problems in terms of their importance and to find the most elegant type of solution to them. Nobel Prize winners prepare their apprentices to be part of the scientific community, but also to be a member of their elite. They notably impress upon them the extent to which this elite can be rigorous when it comes to assessing their own work and that of their peers. Researchers who have undergone this socialisation process gain self-confidence and rapidly access important roles (Zuckerman, 1977). The Merton–Zuckerman couple were able to put this hypothesis to the test through their own personal experience; their son obtained the Nobel Prize for Economics in 1997.

Research directors play a key role in initiating the process of accumulating advantages for their students (Long and McGinnis, 1985). Wagner (2006), looking at the careers of physicists and violin virtuosos, shows that the effects of reputation rely on a coupling between master and disciple. With the notion of 'coupled careers', she questions the usual perception of careers as being individual and based on personal work and aptitudes that escape sociological analysis. The notion designates a phenomenon of interactions between the related careers of two researchers. The reputation of one contributes to that of the other: being a pupil or a former pupil of such and such a renowned scientist, or having co-signed articles with such a scientist, helps young researchers to acquire fame; conversely, supervising brilliant pupils (the best PhD students) is a means for senior researchers to garner esteem for themselves. In the case of coupled careers, the Matthew effect acts on two people at the same time. This phenomenon can also be observed in some male–female researcher couples, while in others the success of one (often the man) is to the detriment of the other.

The Matilda effect: theory of cumulative disadvantages

The Matthew effect leads to greater recognition than deserved. The 'Matilda effect'[2] describes the opposite phenomenon and affects female researchers (Rossiter, 1993). This notion reflects a process of cumulative disadvantages. Female researchers are subjected to the unequal structure of the world of science. Universities delayed in opening their doors to them (1971 for Princeton). Throughout their professional career, women receive less incentive, support and recognition. Their directors (often women) are usually less prestigious; male research directors tend to lose interest in these female researchers (Long, 1990). When they do a post-doctorate, women build up less of a reputation than their male counterparts (Reskin, 1976). They are not recruited as often in prestigious institutions and their access to the financial, technical and human resources needed to achieve scientific results and capitalise on these at international level is more limited. They very infrequently become team leaders, considered to be an essential step in a scientist's career. This phenomenon of cumulative disadvantages affects women and ethnic groups especially. Those who are not ejected, in the name of quality standards, in reality develop a sexless or androgynous personality, a phenomenon that is seen to confirm the sexism of scientific institutions. The Matilda effect is even more blatant given that, although women researchers are less prolific, the quality of their publications surpasses that of men's in terms of how many times an article is cited (Long, 1992). Female researchers are faced with a glass ceiling.

For a long time, the lesser career status of women was explained by their low level of scientific productivity. Being less productive, they were unable to stock up on the necessary signs of academic recognition. It was argued that this was due to a dissymmetry in family chore-sharing, which prevented women from being devoted to science as they should. Zuckerman et al. (1991) contested this explanation showing that men and women overestimate the incompatibility of family chores with intense scientific work. In fact, male researchers with families publish more than those without, while there is no difference in productivity between women according to whether they have children or not. Being married with children has no significant impact on female researchers' productivity (Xie

and Shauman, 2003). However, the disparities between male and female researchers as to the positions occupied and salaries earned continue to exist. In the
United States medical field, although 40 per cent of students are women, only
20 per cent of doctors, 5 per cent of department heads and 3 per cent of medical
faculty deans are women. In France, in 2002, 38 per cent of lecturers and only 15
per cent of professors were women. At the CNRS, they represent 37 per cent of
research fellows, 24 per cent of 2nd class research directors, 12.4 per cent of 1st
class research directors and 6.8 per cent of exceptional research directors.

A decisive moment in a researcher's career is achieving the status of research
director. However, the appraisal and selection criteria implemented by 'peers' (75
per cent of these being men) are decidedly masculine. The model is that of 'the
big firm manager with lots of potential': future lab directors, up-and-coming
scientists offering personal skills such as dynamism and enthusiasm, destined to
lead 'brilliant careers', be posted abroad, shoulder a wide range of responsibilities (in terms of teams, technical platforms and various commissions), and add
their name to publications hence lending them greater visibility. The age barrier is
especially fatal: at age 50, researchers (often women), who have not been unworthy, are nevertheless passed by when there are promotions.[3] It would appear
that the barrier encountered on the career ladder is the result of this progressive
accumulation of small discriminations over the years (Marry and Jonas, 2005).

Throughout their career, researchers are subject to appraisal, including when
they apply for resources. Loyalty, confidence and compatibility of intellectual
styles all have an influence on such appraisals, often without the appraisers themselves realising it. The latter sincerely believe that they are acting in an objective
and impartial manner. Appraisals would appear to be carried out in a relatively
transparent and consensual manner, leaving little room for the patent discrimination of women. However, while equal treatment can be observed when it comes to
recruiting lecturers, this disappears when professorships are at stake.

The Elite and the Mass

The system of recognition explains that some researchers are highly productive.
So what exactly motivates others who have a low level of productivity and obtain
little recognition? Are they impelled by the impression that their modest contribution helps to move science forward? What are the respective roles of the elite
and the mass of researchers? The Lotka and Price laws suggest that the dynamics of the sciences can be explained by this productive minority. The 'Ortega
hypothesis', on the other hand, suggests that the contribution of the mass of
science workers should also be taken into account: 'The majority of scientists help
with the general progress of science even if they are walled up inside the narrow
confines of their laboratory' (Ortega y Gasset, 1932, p. 84).

According to this theory, elite researchers are able to make their contribution to science thanks to the work of the scientific masses. Looking at political
sciences, Dogan (2000) identifies 20,000 researchers cited in the Social Science
Citation Index. Out of these, two thousand are each cited 200 times. These two

thousand constitute the elite, or the high flyers, who stock up a total of 400,000 citations between them. They can be broken down into three categories: those whose work is located at the intersection of several disciplines; those who were educated in the most prestigious breeding grounds; and those who are pioneers in their field. The 18,000 other researchers are cited on average 40 times during the same period. Together they account for 720,000 citations, that is, twice as many as the elite. Their production directly conditions the success of elite scientists, working at the intersection of several fields, and pioneering scientists, whose fate depends on the mass of their followers.

Exchange and the Social Link in Sciences

The mechanisms behind the social sets making up the sciences also reflect the type of social link between scientists. In Merton's thinking, norms govern individual behaviours. His model is based on economics. Scientists are individuals who compete to produce objective knowledge in order to acquire esteem.

Gift Giving and the Social Link

Warren O. Hagstrom (1965) puts forward another model to explain the social link in sciences. For him, social control comes from a system of exchanges and reciprocity between scientists and not from a system of norms. The institution of science is a space for the movement of goods. While Merton's model is quasi-economic – scientists compete to take priority over discoveries – and quasi-legal, although devoid of either a legislative or a policing body, Hagstrom puts forward a pre-capitalist economic model, as described by anthropologists (Figure 5.2).

His survey on scientists from the best American universities reveals that they are all searching for recognition. They are less motivated by a disinterested concern to make scientific progress than by a keenness to extend their personal recognition. This is not a mechanism encouraging them to comply with norms but a driver of individual dynamics. Researchers above all produce knowledge to gain esteem and speed up their own personal advancement. In return for the knowledge produced and in order to obtain this knowledge, the institution bestows

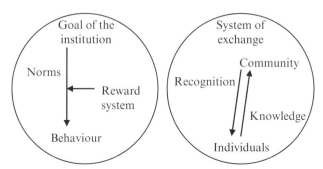

Figure 5.2 *The institution of science according to Merton and Hagstrom*

upon them recognition. A community of interests is created between researchers and institution. Each has an interest in trading the goods they can offer.

Scientific exchanges are very different from the trading of commercial goods; scientists do not sell their work. They offer up their output as a gift to the entire community, which, in return, offers them recognition: they exchange gifts. The goods traded are done so free of charge. A subsidy, for example, is not a commercial transaction, but a gift bestowed on researchers to encourage them to continue their efforts. Contracts specify that researchers have an obligation of best intents, that is, they are expected to do their best. This is more important than the result, which cannot be guaranteed after all. The same applies to grants, equipment, prices and invitations to speak at conferences or appointments to research or teaching positions. It is not a question of repaying past merit. It is about the scientific community giving a gift: peers award researchers with resources, in line with their existing contributions to the community, and the researchers thus have further opportunities to shine. Conversely, scientists who submit an article for publication make a contribution to a review. In the case of some specialities, this contribution also requires supporting the review financially (by subscribing to it or by financing the publication) as well as writing in it. When researchers teach, they *give* a lesson or a seminar; they are not paid for their discoveries in the process.

The fact that nations and companies destroy part of their resources, without any obvious return, can be compared with an immense gift, like a potlatch or the public and ostensible destruction of wealth. Thill (1973) thus describes the particle physics laboratory and experiment as a 'scientific party', a parenthesis in the usual routine where wealth that has been carefully stored up is wasted in one go. The destruction of wealth is ostensible and is a challenge for rivals to dare to do as much. Nations and companies compare each other in relation to the amount of expenditure agreed for research (percentage of GDP or turnover). These gifts are not given out of interest. In the very act of gift giving, disinterest is displayed:

> 'The thing received is disdained, it awakens defiance [organised scepticism], and is only picked up for a moment after being thrown to the feet [as a sign of disinterest?]; the giver affects an exaggeratedly modest attitude [the legendary humility of the scientist?]; after having solemnly, and ceremoniously, produced his gift [in accordance with the norms of communality?], he appologises for only being able to offer this mere scrap [again reflecting scientific modesty: 'I am but a dwarf standing on the shoulders of giants'], and throws before his rival and partner the thing given [as an ultimate gesture of pride and self-importance?]. However, whether discretely or with loud trumpetings [publication or oral communication of discoveries?] the solemnity of the handover is proclaimed to all [the official claim to the paternity of the publicised discovery and public notoriety as a counter-gift?]'. (Ouellet, 1987, pp. 119–20)

Widespread Exchange and Total Social Fact

Gift giving is disinterested but carries with it obligations, according to Marcel Mauss. Although apparently free of strings, a gift is actually binding (owing to past gifts received) and given with an underlying interest (in terms of future gifts). A gift creates asymmetry between the giver and the receiver. Accepting a gift implies a form of recognition with respect to the giver, like a debt that is owed. The object of the gift is always marked with the name of the giver (that is, the scientific contribution is signed), unlike goods sold on the market. It therefore represents the giver. The gift creates a reciprocal duty; it requires a counter-gift (which can be neither equivalent nor automatic as this would be tantamount to sending the gift back to the giver, hence undermining the entire process). This reciprocal duty is not towards the giver, but towards the group as a whole. The receiving party must give in return, but not necessarily to the giver. The most important thing is for the receiver to give something in exchange. If the group is small and each member gives to the others, then each receives a gift. Gifts attract counter-gifts. The exchange is circular within the group. When a scientific review accepts a researcher's contribution, it recognises the author's superiority and strengthens his/her status. When several reviews accept a researcher's contributions, the scientific community develops a debt towards that researcher. Those who contribute substantially, make several discoveries or produce significant writings, give lessons and lectures, train colleagues and supervise their work, or take time to sit on different committees, oblige their peers to respect and recognise them. Their peers have to give in return. The system is based on a threefold obligation: (i) scientists give the knowledge they produce; (ii) their colleagues receive this knowledge and; (iii) in turn, they give the scientists esteem, resources and new knowledge.

The science exchange system covers everything: money, capital goods (equipment, samples and so on), information (publications, pre-publications, data and so on), people (visits, invitations, appointments and so on), prestige, renown, credit, authority and visibility. It is a 'total social fact', as Mauss would say.

This free gift giving, with its attached obligations, is the central mechanism in the science exchange system from which a community is built up (compulsory reciprocity) along with its values (selfishness/unselfishness, pride/humility, courtesy – reading each other's work – and politeness – swapping marks of esteem). The results are given in exchange for specific rewards within an integrated community according to gift-giving rules. Gift giving forms the basis of the normative system described by Merton.

The work of Hagstrom is a turning point in the sociology of the sciences owing to its insistence on competition for recognition and the norms of independence and individualism. It led Merton to rework his first analyses with the concept of ambivalence while others examined systems of reward, allocation of status and stratification.

Social Control

While researchers compete for notoriety, the community knows how to bide its time. It hesitates before recommending or reprimanding a group or a new theory. It uses its gift-giving system to regulate turbulent activity. It puts the reins on controversies by offering them frameworks. It returns overly arrogant, personal, ambitious or controversial texts to their authors. The gift-giving system is a mutual monitoring mechanism. Its operation is made smoother by the fact that its members imagine that they are acting spontaneously, in accordance with their personal tastes for a given subject. In reality, their desire for recognition makes them worry about what their peers think of them. Obtaining peer approval is at the core of the social regulation of the sciences. Researchers conform to the norms of the institution because it is in their individual interest to gain the approval of their peers. They worry about what they give (quality and originality of their work), more than if they were paid for the service. Because the gift, which is to be viewed and judged by their peers, is part of them to a certain extent, they feel obliged to be careful about what they offer. Social control is anticipated. For the same reasons, scientists become conformists, push aside any revolutionary tendencies and take a wealth of precautions, notably in terms of what they say, so as not to rub anyone up the wrong way.[4] The gift-giving system has a political function in that it pacifies the scientific community in relation to the potential violence that reigns within. This violence comes from researchers striving to challenge the work of their peers. Researchers are at once conformists, with respect to their peers, and revolutionaries, with respect to society and acquired knowledge.

Competitive Struggles for Scientific Credit

The exchange model was taken on board by the economy of the sciences (with the notions of monopoly, information asymmetry, moral risk and anticipation) and rethought by Bourdieu (1975). He saw within science a social field, an arena of competitive fighting to build up a symbolic capital, or scientific credit, and to gain control of authority in science. He reproached Merton for being a victim of the professional ideology of scientific circles and for not seeing the reality of the founding principle of this so-called 'community', that is, a market of symbolic goods where scientists spar with each other to maximise their symbolic profit. Bourdieu places the notion of symbolic capital in the centre of his explanatory model. Unlike the notion of recognition,[5] *credit* is possessed, hoarded up and invested in. Making it possible to distinguish between those who have credit from those who have not, this notion leads into an analysis in terms of social classes. As for scientific knowledge, this is a resource that the researcher swaps on a sort of market in exchange for scientific credit. This will then be invested in research areas so as to maximise profit on the competitive market of scientific productions and gain further credit. Researchers aim above all to accumulate this symbolic capital just like capitalists accumulate money.[6]

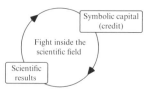

Figure 5.3 *Scientific credit accumulation cycle*

Knowledge Value and Scientific Authority

According to Bourdieu, goods (that is, scientific knowledge) do not have an intrinsic value in science. Their value comes from the fact that they can be exchanged against other goods. This value depends on the importance that the others give to the thing that is exchanged. Scientific output does not therefore draw its value from its compliance with reality or with scientific and ethical norms, but from the interest that other researchers have in it and from what they are willing to give in exchange. It follows that researchers choose their research subjects and areas in relation to the probable importance given to them by their peers. This scientific production value can vary over time (it can be appreciated and then rejected, and vice versa), and in space (it can seduce the British while irritating researchers on the continent, or be welcomed differently according to disciplines or schools of thought). There is no overseeing and neutral authority able to attribute value besides the balance of forces between scientific groups with competing interests. Researchers must therefore fight for the recognition of this value, for example by imposing new appraisal criteria: norms and rules pertaining to its scientific nature. Based on a rhetorical analysis of a publication by Guillemin, Latour and Fabbri (2000) show how this researcher redefines the technical criteria applying to research in their field. The fact that they succeed in having their new standard accepted enables them to deflate the claims of those already holding the results, but who had obtained them using different methods.

Competitors are quick to use their scepticism and deflate the claims of an agent when it comes to originality and scientific nature. Furthermore, the value of scientific output fluctuates according to the transactions effected in the specific field. The value is determined at the time of the exchange and by the exchange. It is normally never a given. Researchers must therefore handle the fluctuation of these values and develop appropriate anticipation strategies. Scientists behave as capitalists. They attempt to place their products at the right time and in the right place in the scientific field, by investing in the most profitable subjects and methods in relation to demand. They must also be familiar with the state of the field (system of objective relations stemming from previous struggles). They can then exchange these scientific values for social values.

Scientific credit is equivalent to symbolic capital acquired by agents; it is recognisable by visible signs such as titles, prizes received, positions occupied and responsibility for equipment. It is made up of scientific authority or competency, that is, an indissociable mixture of technical ability and social power; judgements concerning scientific competency are always influenced by knowledge of the position occupied in the field. Agents hold unequal amounts of credit. The initial capital plays a determining role. The trajectory of agents is conditioned by their previous path and by their current position in the field. Scientific authority is the ability both to speak and to act in an authorised and authoritative manner when it comes to scientific matters. It is the result of social recognition and makes action in the scientific field possible.

Scientific Strategies and Field Structuring

Scientific credit is earned through struggles to have the value of a theory admitted. These struggles are not merely cognitive; they also act on the definition of the scientific field and the directions in which it is heading. The quest for credit involves domination and monopolisation strategies directed against other agents in the field. Agents determine their strategies in relation to the anticipated profit. A strategy can consist in investing in a popular, well-resourced field where the stakes have already been defined and the methods tried and tested, where the public is just waiting to pounce on the slightest amount of progress and where there are active and well-organised exchange networks enabling discoveries to be quickly escalated. Nevertheless, in such a field, competition is fierce. This is why some prefer to adopt other strategies, for example, by investing in less popular fields, where it is easier to find a footing and acquire a monopolistic position. Agents can also invest in a marginal field, potentially promising in the long term. Still others, like Guillemin and Andrew Schally, put in huge amounts of effort, in terms of time and money, in a very popular field, where the stakes have been well defined but where the methods have not yet been fully developed. In spite of their early failures (14 years of unfruitful research) and difficulties, these researchers persevered knowing that if they did succeed, the financial benefits (therapeutic usefulness) and scientific rewards (Nobel Prize) would be huge. Theirs was a long-term strategy, requiring such heavy investments that competitors were discouraged. Thus, the search for credit involves making strategic choices (field, method, place of publication and so on), considered as investments from which agents can draw a maximum amount of profit.

Given that scientific fields are relatively closed, competitors form a sort of community within them. This is defined by the common features shared by community members: values, beliefs, practices and *habitus*. The *habitus* is made up of the set of rules that are learnt and adopted by the agents in a given field. It reflects their past experience (structured structure) and defines future attitudes and behaviours (structuring structure). For a given scientist, attitudes and behaviours are defined by their *habitus* (that is, that of their community) and by the position occupied in their scientific field. Young researchers struggle to get their names to appear first, hence ensuring better visibility, while recognised scientists willingly accept to having their name appear second, given that they will be noticed anyway (Zuckerman, 1968). This behaviour is due less to a norm of generosity than to the opportunity to be seen as magnanimous towards a colleague and hence strengthen a position of superiority. Bourdieu's analysis can be used to reinterpret Merton's norms: scepticism is a means of fighting against competitors and undermining the value of their output; humility is a way of emphasising greatness; communalism is a necessity imposed by the scientific production system in which agents need the work of others. Publication and information exchanges are less dictated by the norms of politeness and courtesy than by the need to improve productivity and be familiar with the state of the field and competitors' strategies. As for disinterestedness, this is simply a rhetorical strategy dissimulating an overriding motive to dominate the field.

These struggles result in the scientific field being structured between dominators (those with the credit) and dominated. The dominators fight to maintain and reproduce the established order; the dominated are forced to use ruse and subversive strategies to attempt to modify the balance of power. Any action, even cognitive action, performed by a scientific agent stems from a strategy to increase his/her domination of the reference field. Agents' arguments and justifications are dictated by their position within the field; they are driven by forces beyond their control. The production of new, valid knowledge is the result of scientific agents competing with one another and exerting mutual control over one another.

Credibility Cycles and their Extension

Latour and Woolgar (1979 [1986]) took Bourdieu's model and proposed a version of it that took into account the diversity of working practices and the multitude of actors' motives. They queried the reproduction of symbolic capital. They challenged the analysis of the credit accumulation system owing to its one-dimensional aspect and the fact that it does not take into account the content of scientific production. They suggested replacing the notion of 'credit' with that of 'credibility'. They claimed that sociologists, from Merton to Bourdieu, had been mistaken about honorific rewards. Furthermore, they argued that recognition requires other, more tangible forms (grants, posts) that are not only the 'visible signs of symbolic capital', but also resources for activity. It is therefore unnecessary for scientists to achieve honorific awards in order to follow a brilliant career, access the right jobs and receive the right kind of gifts to be able to work. While the notion of credit reflects the scientific authority of the elite, credibility considers the fact that other researchers are able to continue to work.

The advantage of the notion of credibility is that it also mixes internal and external factors, scientific data, instruments and recognition, as do the researchers themselves. It applies to strategies where the aim is to invest in equipment, hypotheses, colleagues, publications and places. This investment pays off through publications,[7] but also through young, well-trained researchers and well-liked equipment. This output is converted into a 'credibility cycle' in the form of peer recognition and new resources to be invested in order to maintain and develop the activity (Figure 5.4).

Thus, recognition would only be a secondary phenomenon if it could not be converted into something else. The extension of the credibility system is more important for scientists than the truth, data, research topic, publication or recognition. Researchers' productivity in terms of publications, for example, depends on the scope of their credibility cycle. This increases throughout their career as they spend less time on research and more on teaching and administrative tasks (for example, sitting on the management committees of scientific institutions or the editorial committee of a review). Supervising the work of several researchers

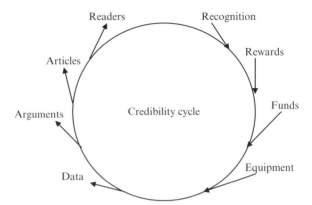

Figure 5.4 *Credibility cycle*

means that they can co-sign their publications without actually doing any of the time-consuming groundwork. Contributing to the definition of national research programmes provides them with an overall view of the field. Researchers' performance above all reflects their ability to get involved in several projects and extend the credibility cycle they control. This ability depends on the structuring of their field, for example, the splitting up of tasks between theoreticians and experimenters, or on how mature the field is.

Extending Credibility Cycles to Society

The credibility cycle moves from the hard-grafting researchers to the referees sitting on review boards, editorial boards and research committees. Rip (1988) suggests extending this analysis, operated at micro level (laboratory, resources deployed and output), to the meso level (the layer of institutions in charge of organising scientific activity, public research councils and programmes) and the macro-level (public legitimisation of science, definition of political objectives and research missions). These three levels interact with one another. Institutions put constraints on researchers and influence their orientations. They do this by allocating resources in relation to their thematic priorities, while at the same time providing the researchers with resources and legitimate arguments for them to pursue their projects. Researchers' output is a resource at the meso level: it helps to strengthen the researcher; the researcher's results become the results of the supporting university or programme. They help to transform the field, by creating new priorities for instance. Similarly, they affect the societal context (awareness-raising, new thinking, new objects and so on).

Rip suggests taking the credibility cycle notion and applying it, for example, to the way a research council operates (Figures 5.5). Like researchers and laboratories, research councils have to earn their budget by showing governments and the public that they do worthwhile things with the money they are given. They need positive campaigning and push researchers to acquire this from their peers (by doing a lot of publishing and mentioning the name of the research council

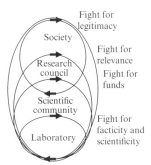

Figure 5.5   *Extension of credibility cycles*

that has allocated the subsidy). Research councils and laboratories depend on each other in their efforts to obtain financial resources. While researchers struggle to prove the facts and the scientific nature of their work, laboratories and research councils struggle to obtain financing. To obtain funds, they have to allow their backers (citizens, elected officials and economic actors) to judge the societal relevance of the projects proposed. The credibility of large programmes (spatial, fight against cancer, nanotechnologies) and research agencies depends on the relevance of the research projects they support. They are thus engaged in a struggle to ensure that their choices are found to be relevant. Moreover, the way research programmes and researchers adjust their projects in order to mobilise each other has been shown through surveys (Rip & Nederhof, 1985).

Over the last few decades, new balances have been set up through scientists' interventions in the media and public debates. These reflect their struggles to have the legitimacy of their scientific research recognised. Researchers' movements (for example, the French 'Let's save research' movement) and their unions, born with the creation of large research laboratories (at the end of the twentieth century), are also engaged in these battles to have their social legitimacy recognised. Although these movements use the union rhetoric of workers denouncing the exploitation of their work by laboratory heads, what they really aim to do is to have their social identity recognised along with their status as researchers. One might think that the scientific elite do not encounter such problems of identity. However, this is not the case: if the scientific elite were a social elite they would have the authority to determine the root of the problems facing society and how best to solve them (Weingart, 1982). In fact, such authority is wielded by hybrid elites (made up of scientists, industrialists, financiers, politicians and others) rather than purely scientific elites. Examining credibility cycles in science brings to light several ways to analyse the intertwined relations between the sciences and societies.

Social Networks in the Sciences

Hagstrom's focus on gift giving underlines that scientists are beings in relationships. They offer up their work to their peers; they have the courtesy to read one another's work and hold one another in mutual esteem. As competitors, they keep an eye on one another. However, there are many others who still think of science as an area populated by isolated individuals. Bernal (1939), for example, who established links between science, ideology and economic infrastructure in accordance with the historic materialism approach, presents scientists as isolated beings who only know of their contemporaries via books and work making them famous. Ben-David and Collins (1966) explain the emergence of new specialities as an individual migration phenomenon, but never as the transformation of a web of relations.

Yet, researchers work in teams, within laboratories, organisations and networks of scientific cooperation, where their work is coordinated. They develop relationships based on cooperation, and on hierarchical orders and affinities. They disseminate their texts to select colleagues before publishing them. This allows them to garner pertinent remarks, improve the text or anticipate possible objections. They write to one another, call one another and meet up to discuss their work. They develop their own networks of relations.

Personal Social Networks

The notion of 'personal social network' whose use converges with the theory of social capital inspired by Bourdieu (Burt, 1992), characterises the 'resources embedded within a social structure'. Individuals have access to these resources and use them to pursue their objectives. A researcher's personal social network varies according to the laboratory's social hierarchy (Box 5.2).

A personal social network is characterised by its size, the type and the diversity of people communicating with one another, the type and force of relations, and the density of relations between network members. The network constitutes a resource (social capital), but also *a* constraint. Its building and maintenance require time. Furthermore, engaging in relations creates proximity, which in turn increases mutual influence. The weight or pressure of the network can thus be so great that they affect an individual's creativity. The weak and non-redundant links allow innovative ideas to spread better, but do not provide access to the strategic resources of network members.

Social Networks in the Sciences

A personal social network is an individual's network, but sociological analyses also focus on sets of individuals and their relations with one another. The sociometry of the 1930s opened up new perspectives in this field, but it was not until the 1960s that authors began to consider these characteristics in relation to the study of the sciences.

Invisible college
The physicist and science historian Derek de Solla Price (1963) studied the dynamics of the sciences using publications. For him, the article is an objectivated, classified, dated and quantifiable form of scientific work. Studying large

Box 5.2 *Different networks according to the echelons (Shinn, 1988)*

A research director's network is far-reaching. Directors devote half of their time to communicating with other scientists inside and outside the laboratory. They are involved in various commissions, take part in talks with researchers holding key positions, dialogue with other laboratory directors, sit on PhD examining boards, participate in the design of science museums and act as consultants in companies. They are in regular contact with dozens of researchers, administrators and users of science. Their network can include up to 200 people and extends across the globe. It enables them to exchange scientific information and access data and new ideas that they then pass on to their laboratory. Their network takes up a lot of their time but does not disadvantage their intellectual productivity; they benefit from the results of their researchers and link these up with other results to which they have access via their network. They are informed about what is being sought and practised outside of their laboratory. They sign a high number of publications.

Young researchers' social networks, on the other hand, are limited to around 20 other researchers along with several senior researchers. They use their networks above all to exchange experimental data, information about instruments and conceptual representations. Sometimes they are in contact with an instrument seller or a technician, or enter into contact with a neighbouring laboratory in order to procure an instrument or a sample.

Senior researchers have broader networks, including above all other senior researchers in different laboratories. They are involved in networks that they use occasionally for personal or scientific reasons. Sometimes, they are in contact with research administrators to obtain credit, or with non-scientists for teaching or popularisation purposes. These networks enable them to extend their ability to analyse and disseminate their results.

sets of publications, he built up temporal series and drew some general lessons: scientometry (of which he was an inspirer) works on volumes of publications and their relations: productivity and influence-related phenomena. He assumed that science progresses according to a logistic curve (S-curve) with a preliminary period (few publications, low growth), an exponential growth phase (the number of publications doubling every two years), a saturation phase (constant number of new publications) and a decline phase (until the field has been dissolved). It is therefore possible to follow changes in a laboratory, field, country or discipline: structuring of a community or emergence of specialities, influence of an international review, relations between science and technology.

An article is not only a piece of information that refers to an author or content; it is also the expression of a form of social unity that can be seen when the references cited in a text are examined. Science progresses thanks to former output,[8] on which new articles rely in order to make their own contribution. Articles are cited and therefore connected: it is possible to analyse their relations. Price differentiates between two types of citation: (i) archival citations, from past literature, corresponding to dated or precursor texts; and (ii) 'research front' citations reflecting the relations of proximity between contemporary researchers. Citations make it possible to sketch the outlines of social groups, especially since researchers are aware of the close distances between one another and therefore tend to cite one another: they form 'invisible colleges' that have no institutional visibility as groups. However, their members are visible since they are cited and recognised in their institution. They often know one another, have met at conferences, and may even have cooperated. They develop strategic synergies allowing them to control certain directions of their scientific community and local institutions. They constitute a power group within a speciality. These colleges are small to allow functional communication; above one hundred, the group tends to spontaneously subdivide itself. In its midst, a distinction develops between a 'clique of leaders' and the ordinary members. Basing his study on the highest citation indices, Price above all observed the elite.

Social circles

Crane (1969 and 1972) studied the spread of scientific innovations in agriculture and gave an empirical basis to the invisible college idea. However, she preferred the notion of 'social circle'[9] to underline the importance of informal relations between scientists, whatever their visibility. She showed that researchers do not need to know one another personally to influence one another. They can exert their influence via intermediaries. She questioned researchers about their relations and asked them with whom they kept in regular contact. She showed that relations are organised around a small number of 'opinion leaders', often the oldest and most visible. The group then experiences growth (via the recruitment of students) and a process of breaking up into smaller groups (30 members maximum). The communication between these smaller groups deteriorates, innovations are disseminated less and receptivity to the outside world decreases. The building of the scientific elite can also be explained through the analysis of networks.

Network mapping

Analysing social networks makes it possible to establish maps of relations and to analyse the flows of information, contacts between scientists and influence. Relations are identified based on questions about regular contacts, meetings, collaboration, perceived influences and resource people. The data come from interviews, surveys, questionnaires, monitoring of exchanges (for example, the analysis of correspondence) and relations between publications (or patents): co-signatures, citations and co-citations.[10] Graph analysis is also of instrumental use in the sociology of the sciences (Callon et al., 1986). Maps of relations are drawn and indicators put forward. Using mapping it is possible to monitor the changes to and structuring of a field, taking into account phenomena such as self-citations, cross-citations (between authors returning a service) and 'salami-science' (the breaking up of results across different publications).

Indicators can be used to characterise graphs of relations: (i) *density* measures the relative frequency of relations between entities. When a set of points shows a high density of relations, it can be isolated and seen as a group with defined borders (cluster, spin-off or aggregate). A low density reveals a set with loose and dispersed relations.[11] (ii) *centrality* measures the relations of one entity with the rest of the network. It quantifies the importance of this entity for the others. The entity can be central or marginal.

Scientific networks do not necessarily correspond to divisions between disciplines or specialities. The members of a network do not necessarily share common characteristics as one might expect according to a logical classification. They are heterogeneous. Furthermore, the divisions are not clear divisions; researchers belong to several networks. Almost 50 per cent of chemists say that they have research interests in more than one speciality (Blume and Sinclair, 1974). Moreover, the links and composition of a network are perceived differently by each participant; it is therefore difficult for an outside observer to find an objective criterion to define a network in an unambiguous manner (Woolgar, 1976).

Network dynamics

Relations between publications constantly change. Scientometry observes an immediateness effect arising from the fact that recent publications are more often cited than older ones. This is measured by calculating the citation half-life of an article (which supposes that a publication has a life cycle over the course of which there is a decrease in citations). It varies according to disciplines: in life sciences it is a 3-year cycle. Some publications are not affected by this phenomenon and are increasingly cited over time.

Networks are fluid social arrangements; they are born, grow and then die. Their composition changes over the course of time (Mulkay et al., 1975; Geison, 1981). By combining a structural analysis with a longitudinal analysis, scientometry characterises their transformations: budding, fusion, fission, building of obligatory points of passage, emergence of a speciality, and internal densification of a group or loosening of relations. It identifies the fields with the highest level of activity: notion of research front made up of around 50 or so articles having an immediate effect. Once they appear, these articles lead to summary and critical appraisal work. Scientometry also identifies the central corpus of founding texts, similar to the corpus of myths in ancestral tribes, with its core of scientists forming the pillar in a field. Identifying research fronts, founding cores and networks helps to delimit a social group, a field or a scientific problem. Identifying one helps to define another and vice versa.

The combination of density and centrality indicators makes it possible to position fields in relation to each other in a diagram (Figure 5.6), and to see how these positions evolve. Callon et al. (1986) observed a majority of situations in which the field emerges as something marginal or loose and then becomes denser; its members move closer to each other and are involved in an increasing amount of exchanges. The field is still marginal at this point. Then comes a phase during which the field is recognised and becomes even denser (internal consistency) and central (recognition). It can also disappear, as it is absorbed by other fields (remaining central but on a looser basis) or become loose and marginal once more.

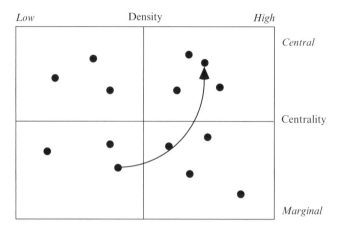

Figure 5.6   *Diagram of the positioning and changes in fields/networks*

Network analysis makes it possible to study the movement of cognitive influences, both direct and indirect, and the long networks that are built up in a field or from one group to the next. Granovetter (1973) showed that weak links pass on influences over longer distances and between groups with few connections; these explain the privileged role of marginal individuals in the spread of new ideas.

Network heterogeneity
By extending the credibility cycle notion, Rip (1988) introduces a whole series of actors (financial organisations, the state, private partners, the media and citizens), hence broadening the scope for analysis. Science is not a closed space. The production of knowledge is marked by multiple interactions with other actors in society. Science networks are heterogeneous; they are composed of researchers from different specialities[12] and non-scientists. What they produce is negotiated; it gives rise to discussions between actors who are not only scientists. For a given project (for example, the development of nano tubes), the relevant analysis entity is neither the laboratory nor the scientific speciality but a heterogeneous network composed of actors from public research, companies and the state. When such a heterogeneous network is highly interactive and consistent, it tends to behave like a single actor, referred to by Callon et al. (1986) as an 'actor-network'. When Guillemin set up his laboratory, he brought together physiologists, neuro-endocrinologists, chemists and biochemists. He also involved university administrators, financial organisations (National Institute of Health), pharmaceutical companies (interested in the growth hormone release factors), large slaughterhouses (supplying large quantities of hypothalamus) and foundations. The laboratory was hence nothing more than a node within a network of laboratories, companies, hospitals and patent offices.

Research work is made up of relations and activities that pass through the laboratory.[13] This contextual dimension of scientific work can be seen by the observer in different ways. Researchers write letters and send off draft articles and research proposals; they make phone calls and leave for visits and meetings; on their return, they explain what has been said during these meetings. They can thus be seen modifying their research proposals, rewriting their articles and reworking the terms of a contract signed with a company. They can be seen managing the supply of equipment and organising the exchange of samples. Their laboratory work is also structured with respect to their involvement outside of the laboratory. Their commitments and negotiations go beyond the walls of the laboratory and the limits of their speciality. The heterogeneity of their networks stems notably from their need for resources. Analysing their resource relations helps us to understand how they build up and carry out their activity. Resource relations are set up and controlled by the interaction of different actors in contact with one another. While young researchers are a resource for the laboratory that hires them, the laboratory is also a resource for the researcher, in terms of his/her training and career. The relationship is co-built and negotiated between apparently different actors, in the same way as other resource relations.

Globalisation and Territorialisation of the Sciences

In this era of worldwide exchange of information, goods, services and people, the knowledge society seems to be increasingly global. The transport means available make it possible for researchers to move around and meet up more often. The world of the sciences has become a 'Small World' (Lodge, 1985).

Science and communication technologies

The interconnection of networks and the protocols of exchange and navigation on the World Wide Web are changing scientific work conditions. However, the extent to which communication technologies are changing the dynamics of science has yet to be determined.

The analysis of scientific cooperation networks shows how important the concrete methods of exchange between researchers are (Vinck, 1992). Two conventional methods are meetings and publication exchanges. The arrival of fax machines played a role in the setting up of scientific cooperation networks by facilitating coordination between researchers from different countries. The telephone did nothing to help conquer the language barrier, especially when it was answered by a secretary with only very basic notions of English. The increasing popularity of European scientific cooperation networks goes hand in hand with the spread of the fax. Today, it is important to study the stakes behind these electronic communication tools (e-mail, online publications, open archives and distributed calculation). Furthermore, it should be remembered that the Web was invented by researchers at the CERN European Council for Nuclear Research to meet their scientific coordination needs.

Scientific districts and proximity effects

The effects of proximity between researchers can be observed at urban level (conurbations, employment areas and so on), rather than at campus or regional level. Studying the case of 'science parks', Grossetti (1995) showed that the relations between scientific clusters in the same region are not very significant. Economic theory explains the proximity effect through communication requirements and the need for personal interactions to exchange tacit or barely formalised technological know-how. Actors have to come closer to one another in order to work together owing to the nature of innovative knowledge. Furthermore, the social network theory decries functionalist and culturalist explanations in terms of local agreements to encourage the setting up of relations. This is why it is indispensable to analyse local networks.

Local social networks foster exchanges between actors, in spite of the barriers separating them (competition, forms of activities and practices, disciplines). They are set up during organised activities (during studies, at work and outside of work), whose concentration and stability condition the importance and density of local relations. Relations are mainly concentrated in urban areas (a transport-hour away); they often stem from belonging to collective entities (families, work, sports clubs and so on), especially those connected with training. They are therefore not the starting point of social dynamics since they are themselves explained through spatial constraints affecting individuals' activity systems. Nevertheless, proximity does not mechanically give rise to local interpersonal networks.

Historical processes involving the local accumulation of industrial and scientific establishments and the harmonisation of these establishments' activities can also be observed. The effects of proximity between these organisations are linked to two time scales: (i) a daily or routine time scale generating proximity effects stemming from diverse social logics; and (ii) the longer, historical time scale of scientific institutions and industrial activities (for example, the 'territorialisation' of engineering sciences in France), whose logic is better explained from the point of view of individuals, their trajectories and their relations.

In this global economy, where science travels across the world via different communication infrastructures, the concentration and agglomeration of scientific resources in a small number of places are phenomena that continue to raise questions. Both public and private research actors tend to group together on defined territories. This leads to the phenomenon of geographic clustering: the setting up of large scientific and technological clusters, sometimes organised for the purpose of achieving economies of scale, pooling of resources (technological platforms), access to a catchment area of highly skilled personnel, access to subcontractors and proximity between people. The ever-growing worldwide communication structures are not in contradiction with this geographically and socially close research.

Market, Mobility and Scientific Careers

In Mertonian sociology, a scientific job and career are components of the recognition bestowed on researchers by their community. Work focusing on the link between merit (peer judgement about scientific excellence) and career promotion pits the authors advocating universalism (Cole and Cole, 1973; Hargens and

Hagstrom, 1982) against those championing particularism and the taking into account of personal attributes and the reputation linked to researchers' academic origins (Allison and Long, 1990). There are nevertheless other approaches to help in understanding issues relating to employment, mobility and careers.

The Employment Market in Research and Universities

Economists have studied scientific employment (recruitment, careers and so on) owing to the atypical nature of this job market. They have focused on the employment management rules and their rationality (Siow, 1998): recourse to the opinion of peers outside of the establishment; the 'up or out rule' that obliges researchers to compete if they want to obtain a position; and lifetime employment for some researchers. They have also looked at the explanation behind the salaries of academics, defined after recruitment without the intervention of review peers. In sociology, the focus is more on how judgements are made[14] and the way decisions to create jobs and recruit and promote scientists are made (judgement market or economy of quality: Karpik, 1989). Appraisals are only partly linked to scientific recognition. Obtaining a position is not the same as being awarded a prize. Universities do not take on professors in order to reward them, but with a view to their future 'output'. It is a question of choosing a candidate who is likely to meet the expectations of the establishment. Their appraisal is thus retrospective and anticipatory; bets are placed on expected behaviour, which explains the preference given to known candidates.

Throughout their career, researchers also undergo other evaluations, corresponding to changes in status ('tenure' in the United States, research director at the CNRS, and university Professor). The conditions for such changes differ according to institutions and national systems. Researchers' professional lives are punctuated by decisive appraisals; these are nevertheless rare events. However, researchers are subject to peer opinion in other testing situations (publications, conferences, lectures and so on). Their career depends on this to varying degrees according to their type of professional belonging and their relationship with the establishment (Musselin, 2005). In France, university professions are relatively autonomous while in Germany, public authorities, spokespersons acting on behalf of the interests of society, have their say in the matter, and, in the United States, scientific professions are influenced by the establishments defining the strategies. The professional model differs according to the country, and is relatively egalitarian in terms of status, relations of subordination and material conditions.

The academic employment market is governed by two employment rules. In the primary segment, employees benefit from advantages (remuneration, stability and career furthering). Economists explain this by the growing specialisation of knowledge and the associated risks for individuals. In life science, the division of labour in laboratories relies on this dualism; employees in charge of experimental work (PhD and post-doctoral students) make up the secondary market segment. In the United States, this dualism has hardened and there is even greater competition to enter the primary market. Some explain this using logic relating to the organisation of research; team leaders invite young researchers to do additional work, because they are less costly than the statutory workforce and are a greater investment because they have to 'prove themselves'. In France, these researchers represent one-third of laboratory staff numbers, some of whom have no salary. There is an implicit contract (Stephan and Levin, 2002) between supervisors and PhD students. The latter undertake to participate in the collective production of the laboratory and in exchange for their efforts their supervisor rewards them by supporting their application for a job within the academic community.

However, the promise is only credible if there really is a job available. In France, PhD students are recruited as permanent members of staff early on, which means that there are few post-doctoral students. Of course, laboratories cannot recruit all of their best researchers. They have to push them towards other laboratories and decide between maintaining skills in-house or mixing them with other researchers (so as to recover the tacit know-how developed in other laboratories) (Hackett, 2005).

In France, at the start of the twenty-first century, roughly 25 per cent of PhD students moved on to a job in industry. This figure could be higher if the students were better prepared but they suffer from a lack of knowledge about the industrial world and its career possibilities. A PhD corresponds to training within a laboratory, that is, professional socialisation within the academic world (acquisition of thinking reflexes, knowledge of the social environment and the way to be and act, building up of networks of relationships and so on). However, this does not prepare students to enter industry, except when there is a partnership agreement with companies. PhD students who are destined to leave the laboratories where they were trained are vectors of the knowledge transfer to other laboratories (Mangematin, 2001).

The Institutional Mobility of Researchers

Scientists' mobility can be of four kinds: *statutory* (promotion, movement within a hierarchy and institutional functions), *institutional* (change of laboratory or institution, move from research to industry), *geographic* (international mobility and scientific migrations) and *thematic* (change of research topic or field). The institutional mobility of researchers during their career is characterised by changes in research focus, instrument or materials used. It is influenced by four factors: the size of the laboratory, the management system and internal hierarchical organisation, the relations between the laboratory and external actors, and the objectives and philosophy of the laboratory's founder.

The question of scientific mobility is subject to controversy: accusations of opposition to progress, and political, managerial and even scientific concerns. The debate focuses on the relevance and feasibility of researchers' mobility. The 'up or out' principle, which consists in moving non-promoted junior scientists to one side, is known to work in the senior scientist selection process while the movement of researchers is supposed to encourage knowledge and scientific creativity. In economic terms, it optimises the innovative effects of discoveries by circulating tacit knowledge (ibid.). However, both employees and employers have difficulty dealing with mobility. It requires costly and risky adaptation of the qualifications acquired and the restructuring of social networks and socialisation in general. Actors are rarely mobile or are reluctant to encourage their best colleagues to be mobile. Before 1969, the CNRS was even worried about the overly high rate of mobility of its researchers (50 per cent) as they moved to universities; opening the possibility of extending contracts, voluntary departures fell to 2 per cent. Managers then became worried about the low level of institutional mobility. The incentives or constraints are linked to the way the academic job markets work and to the formal and informal rules that structure trajectories and professional identities.

The building of an 'impossible' type of mobility: the case of the INRIA
When the French National Institute for Research in Computer Science and Control (INRIA) was created in 1967, mobility was introduced as a career management tool. Three years later, this principle was called into question (Louvel, 2005).[15] The INRIA was a small institute: 350 permanent researchers shared between six computer science, automation and applied maths research centres. This was an original grouping that differed from that of the CNRS and the French universities. Mobility was included in the statutes as a means of strengthening the links between research and industry. The political aim was to catch up on the delay in innovation. In 1966, the government launched the 'Calculation Plan' to support French businesses in the Information Technology (IT) sector, and created the IRIA[16] the following year. The scientists involved in the project strove to institutionalise a discipline that had little recognition in France as well as promote applied research. The organisation was attached to the Ministry for Industrial and Scientific Development, unlike the CNRS which depends on the National Education Ministry. To facilitate the transfer of knowledge, the principle of mobility was included in the statutes as follows:

- The scientists belong to two bodies: a lower body (researchers) and an upper body (research engineers).

- The lower body's contracts are of a fixed term and can be renewed to encourage researchers to move outside of the institute with their results. The IRIA does not provide researchers with a career; it offers a transient period of high-level training in computer science allowing researchers to find outlets in universities, administrative bodies and companies.

- Research engineers supervise researchers and ensure scientific continuity. They benefit from permanent contracts and are recruited for their scientific reputation.

- Some researchers are promoted to the upper body while others leave the organisation after seven years. The statutes set the maximum percentage of research engineers recruited from among the researchers.

This was an original employment management method for the time. It set up a partially closed employment market for the higher-ranking jobs. Rules based on diplomas and seniority set the conditions for access and promotion. In 1970, 20 per cent of the research engineers were recruited from the body of researchers where individuals were only allowed to stay for a seven-year period. The long in-house careers were set aside for engineers whose departure would be prejudicial to the organisation.

This conception of the way the job market worked was based on presuppositions that were out of kilter with the effective employment market. It assumed that: (i) researchers would leave to work in industry of their own accord and would not need any guidance in this (IRIA researchers were familiar with the industrial environment and companies were interested in the knowledge they developed); (ii) the compulsory mobility would operate a soft selection from among the best researchers while the others would spontaneously migrate towards industrial jobs before the end of their contract (hence reducing competition in terms of access to permanent jobs); (iii) researchers are rational and able to assess their relative performance and therefore decide whether to stay or leave; (iv) researchers organise their career in relation to the clear hierarchical order of jobs between the academic sector and the private sector; and (v) the 'up or out' principle used in Anglo-Saxon countries could be applied.

The research employment market, even in the United States, does not operate according to these presuppositions. With the exception of several prestigious universities, over 70 per cent of applicants obtain tenure following a series of selections where few candidates actually apply. At the IRIA, on the other hand, access to the body of research engineers was based on competition between many applicants.

The IRIA statutes were reworked several times. The texts amplified the closed nature of the job market: the percentage of research engineers recruited from among the researchers went up from 20 per cent at the beginning of 1970 to 70 per cent in 1980. The possibility of pursuing a career in the institute was therefore increased. Researchers did not leave of their own accord, which is why their mobility had to be imposed. Every year, a contingent of researchers, who had come to the IRIA to practise research in applied mathematics, came to the end of their contract. Because the number of positions available was limited, the question was what was to become of these researchers? The situation was one of conflict. The competition to get into the body of research engineers was fierce and seen as unbearable.

In fact, the structure of the job market was similar to that of the academic markets. The recruitment channels were not as diverse as expected. The young researchers came from privileged sectors, where the IRIA was used to recruiting. These were promoted by a handful of lecturers and researchers who promoted the IRIA to their students, drawing them in through internships and then recruiting them. Fired by their passion for emerging disciplines, the students were attracted by the theoretical projects developed at the IRIA. The prospect of transferring to industry took second place in their eyes. Research at the IRIA interested them above all because of the academic possibilities that it offered. There was an implicit contract, similar to what can be found in academic environments. The notion of obligatory mobility was foreign to the reciprocal commitments. Instead, the implicit contract could be read as follows: do high-quality research work, which is recognised by your peers, and the institute will reward you by offering you a career. The commitment was personalised, like a form of patronage between a supervisor and a group of pupils. The young researchers collaborated with the private sector, without the contracts pulling them towards industrial jobs. On the contrary, the contracts reassured them in the idea that they would lose out if they tied themselves to an industrial employer, notably in terms of the 'freedom' that a higher salary would have difficulty compensating for.

Furthermore, the transfer of knowledge went well, which did not encourage managers to create opposition to the mobility principle. It used other vectors: contracts, short-term stays that were two-way (between industry and research), theses based on the industrial environment and so on. Having the status of an IRIA researcher allowed these researchers to come and go between different domains. Moreover, a general agreement emerged defining the 'professionality' of the IRIA researchers.

International mobility and scientific migrations

Scientific mobility is also geographic. Researchers move from one laboratory to another, do a stint abroad or have a transfer. This mobility is also linked to the internationalisation of the sciences (Elzinga and Landström, 1996): inter-governmental agreements – the motives of which are not always directly linked to the sciences, like the treaty for the protection of the Antarctica, military cooperation, commercial or diplomatic agreements – which lead to coordinated scientific programmes; foreign researchers can be asked by governments to sit on review boards or scientific councils; and, following the initiatives of some researchers (via scholarly societies or their research organisations), international arrangements are set up and involve the enrolment of the researchers' supervisory authorities. The internationalisation of the sciences is also linked: (i) to the globalisation of trade in goods and services in the economic world, together with the accompanying knowledge (codified and tacit); and (ii) to the types of international division of labour. Scientific output (publications) and technological output (patents) are concentrated in the United States–European Union–Japan triad, which constitutes the main space in which researchers move around.

The migratory movements of highly qualified personnel have increased (Meyer, 2001) under the influence of four factors: (i) the collapse of the Community of Independent States (CIS), which formed a pool of qualified professionals from the Stalinist period; (ii) the rise in school and university education in some developing countries; (iii) the demand for skills generated by the new economy and by the gaps between training and the job market; and (iv) the multiplication of devices designed to ensure the mobility of individuals: acceptance of different national diplomas, worldwide reach of scientific job offers, multiplication of international recruitment agencies and temporary work permits.

Conventionally, these migratory movements have been interpreted within the framework of the brain-drain model, that is, the lasting exodus of skills from the outskirts (Southern countries and Europe) to the centre (Europe and the United States). However, a global movement model would be more appropriate to explain the phenomena of multi-pole and transient mobility observed. In the first model, the migration corresponded to the lasting estrangement of researchers with respect to their country of origin. However, today it is possible to observe episodic returns to the homeland, global commuting in certain regions (between the two sides of the Pacific, that is, the American west coast and Asia) and lasting returns to the country of birth. This does not mean to say that the flow is any less asymmetrical and localised, characterised by pendulum movements, depending on the level of development of the countries.

Out of the 102 million students in the world at the end of the twentieth century, 2 million studied outside of their country of origin. Scientific expatriation differs according to country. Although India and China are developing some impressive diasporas, their expatriates represent only 10 per cent of their highly qualified professionals. For other countries, for example those undergoing crisis, the expatriation rate is much higher. A considerable proportion of expatriates tend to stay in their host country: in 1990, 45 per cent of foreign PhD students in the United States intended to stay there; in 1999, this number had risen to 72 per cent, out of which 50 per cent had already found a job opportunity. However, some do go back home. Between 1986 and 1996, 63 per cent of African PhD students in the United States went back home to professional positions in the scientific and technological fields. In France, students from developing countries account for 12 per cent of the PhD population.

It is against this backdrop of scientific and technological diasporas that a process of knowledge redistribution can be seen to unfold (Barré et al., 2003). Both home and host countries organise their diasporas to enable qualified expatriates to serve the country of origin, while remaining in scientific and technological environments that put their qualifications to use. In the United States in 1999, nearly 400,000 researchers and engineers involved in R&D activities came from Southern countries. These included many Asian expatriates with well-organised networks. Countries like India or China have developed specific policies to send contingents of young qualified individuals to further their learning overseas and 'stock up on grey matter' during their stay before coming back to populate their home countries' scientific and industrial facilities. Dynamic associations, for example the Association of Chinese Biologists, have structured and organised the flows of expatriates. Hence, local scientific communities have joined the scientific research fronts. The rapid development of Taiwanese laboratories, Chinese science parks and Indian start-ups stems from this dynamic movement of scientists. On American campuses, Indian students have invested in the disciplines abandoned by the Americans: mathematics, physics, electronics and engineering.

The above indications point to the existence of complex migratory phenomena that the social sciences are still finding difficult to characterise and explain.

Conclusion: Science as a Regulated Social Space

In this chapter, we have pursued our examination of what might explain the dynamics of the sciences. Several sociological mechanisms would appear to report on the existence and development of the sciences: moral norms, cognitive and methodological norms, and social transactions pertaining to the organisation. Several explanations compete with one another, notably the idea of theoretical moral equality, via the reward system; the idea of specialised communities governed by methodological imperatives and specific organisational resources; the effects emerging from aggregated relations and structural effects. Nevertheless, whether or not these explanations are all equally worthwhile is a different matter. The various approaches have borne fruit, although no one approach in particular seems to stand out from the rest. The question is therefore, do they all have the same explanatory power? This question continues to fuel the science debate. Similarly, there is also the question of the sociologically relevant entity to describe and explain the dynamics of the sciences. There are several possible entities: the community as a whole, specialised subsets, particular organisations, and cross-cutting networks.

Notes

1 Vertical social differentiation, with which the hierarchical organisation of individuals and groups is associated depending on the case, and vertical social division of functions, values and legitimisation mechanisms, or of relations based on dominance.
2 Name given in tribute to Matilda Gage and her book *Woman as Inventor* (1882), which retraced the careers of many famous female inventors.
3 The relationship here with the theory of cumulative advantages and disadvantages seems particularly complex when it comes to researchers' age. Researchers working in the more theoretical disciplines seem to reach a productive peak in their thirties and early forties. This happens later for those specialising in empirical disciplines. There is much debate about the identification of productivity peaks, the rate of growth and declines in productivity, the relationship between age and productivity and the interpretation of the phenomena observed.
4 Hagstrom refers to 'professional rationality', as opposed to the 'contractual rationality' that compares marginal uses.
5 Recognition was defined as a form of reward in a system operating according to the stimulus–response principle and aiming to strengthen the behaviour expected by the institution.
6 Without being quite as radical, Whitley (1974) also founded his analysis on the idea of accumulated reputation.
7 This analysis portrays a contextualistic conception of scientific content which does not have any value in itself but only takes on value through use when it is incorporated into other processes.
8 Hypothesis of continual growth debated in sociology (question of the measurability of science) and in history (questions of scientific breakthroughs and revolutions).
9 Also referred to as 'consistent social group', 'network' or 'cluster' (Price, 1963; Crane, 1972; Mullins, 1972; Griffith and Mullins, 1972).
10 When two texts are cited together in a third text.

11 Mullins (1972) refers to 'network' when the relations are loose and to 'cluster' when they are dense and the group in question establishes its own technical and intellectual norms, manages research content and tends to be institutionalised.

12 Knorr-Cetina (1982) suggests the notions of 'trans-scientific fields' and 'trans-epistemic arenas'.

13 Lemaine et al. (1969) referred to its 'cosmopolitan nature'.

14 For example, the work of Vilkas (1996) on the CNRS national committee, and that of Musselin (1997) on French and German universities.

15 The same phenomenon occurred at the French Atomic Energy Commission's (CEA's) Electronics and Information Technology Laboratory (LETI), where jobs were created for the duration of one year. This period was renewable so that researchers could leave to work in companies with the knowledge they acquired. Today, researchers and engineers come to the institute for the job stability that it offers.

16 This became the INRIA in 1979.

Recommended Reading

References appearing in other chapters: Barber and Hirsch (1962) in Chapter 1; Cole and Cole (1973), Cole (2004) and Zuckerman (1977) in Chapter 2; Stephan and Levin (2002), Vilkas (1996) in Chapter 4; Latour and Woolgar (1979 [1986]), Louvel (2005) in Chapter 7.

Callon, M., Law, J. and Rip, A. (eds) (1986), *Mapping the Dynamics of Science and Technology*, Basingstoke: Macmillan.

Crane, D. (1972), *Invisible Colleges: Diffusion of Knowledge in Scientific Communities*, Chicago IL and London: University of Chicago Press.

Grossetti, M. (1995), *Science, industrie et territoire*, Toulouse: Presses Universitaires du Mirail.

Hagstrom, W.O. (1965), *The Scientific Community*, New York: Basic Books.

Lodge, D. (1985), *Small World*, London: Penguin.

Marry, C. and Jonas I. (2005), 'Chercheuses entre deux passions: l'exemple des biologistes', *Travail, Genre et Sociétés*, **2**(14) 69–88.

Musselin, C. (2005), *Le Marché des universitaires. France, Allemagne, Etats-Unis*, Paris. Presses de Sciences Po.

Price, D. de S. (1963), *Little Science, Big Science*, New York: Columbia University Press.

References

References appearing in other chapters: Bernal (1939) in Chapter 1; Cole (2004), Crane (1969), Lemaine et al. (1969), Weingart (1982) in Chapter 2; Ben-David and Collins (1966), Blume and Sinclair (1974), Geison (1981), Granovetter (1973), Mulkay et al. (1975), Mullins (1972), Whitley (1974) in Chapter 3; Thill (1973), Vinck (1992) in Chapter 7.

Allison, P. and Long, J. (1990), 'Departmental effects on scientific productivity', *American Sociological Review*, **55**, 469–78.

Barré, R., Hernandez, V., Meyer, J.B. and Vinck, D. (2003), *Scientific Diasporas:*

How Can Developing Countries Benefit from Their Expatriate Scientists and Engineers?, Paris: IRD.

Bourdieu, P. (1975), 'The specificity of the scientific field and the social conditions of the progress of reason', *Social Science Information*, **14** (6), 19–47.

Burt, R. (1992), *Structural Holes*, Chicago, IL: University of Chicago Press.

Crane, D. (1965), 'Scientists at major and minor universities: a study of productivity and recognition', *American Sociological Review*, **30**, 699–713.

Dogan, M. (2000), 'Reputation of scholars and scientific specialization', in *Special Session on 'Recent Trends in Political Science'*, Quebec: IPSA, http://www2.hawaii.edu/~fredr/dogan2.html.

Elzinga, A. and Landström, C. (1996), *Internationalism and Science*, London: Taylor Graham.

Griffith, B. and Mullins, N. (1972), 'Coherent social groups in scientific change', *Science*, 177(4053), 959–64.

Hackett, E. (2005), 'Essential tensions. Identity, control, and risk in research', *Social Studies of Science*, **35** (5), 787–826.

Hargens, L. and Hagstrom, W.O. (1982), 'Scientific consensus and academic status attainment patterns', *Sociology of Education*, **55** (4) 183–96.

Karpik, L. (1989), 'L'économie de la qualité', *Revue française de sociologie*, **30** (2), 187–210.

Knorr-Cetina, K. (1982), 'Scientific communities or transepistemic arenas of research? A critique of quasi-economic models of science', *Social Studies of Science*, **12** (1) 101–30.

Latour, B. and Fabbri, P. (2000), 'The rhetoric of science: authority and duty in an article from the exact sciences', *Technostyle*, **16**, 115–34.

Long, J. (1990), 'The origins of sex differences in science', *Social Forces*, **68**, 1297–315.

Long, J. (1992), 'Measures of sex differences in scientific productivity', *Social Forces*, **71**, 159–78.

Long, J. and McGinnis, R. (1985), 'The effect of the mentor on the academic career', *Scientometrics*, **7**, 255–80.

Mangematin, V. (2001), 'Individual careers and collective research: is there a paradox?', *International Journal of Technology Management*, **22**, 670–75.

Merton, R. (1968), 'The Matthew effect in science: the reward and communication systems of science are considered', Science, **159** (3810), 56–63. [This article is reprinted in R.K. Merton (1973), *The Sociology of Science: Theoretical and Empirical Investigations*, Chicago, IL: University of Chicago Press, pp. 439–59.]

Merton, R. (1988), 'The Matthew effect in science, II: cumulative advantage and the symbolism of intellectual property', *Isis*, **79**, 606–23.

Meyer, J.B. (2001), 'Network approach vs. brain drain: lessons from the diaspora', *International Migration Quarterly Issue*, **39** (5), 91–110.

Musselin, C. (1997), 'State/university relations and how to change them: the case of France and Germany', *European Journal of Education*, **32** (2), 145–64.

Ortega y Gasset, J. (1932), *The Revolt of the Masses*, New York: Norton.

Ouellet, P. (dir.) (1987), *Sciences et Cultures. Pour une anthropologie des sciences et des techniques*, Québec: Télé-Université.

Reskin, B. (1976), 'Sex differences in status attainment in science: the case of post-doctoral fellowships', *American Sociological Review*, **41**, 597–612.

Rip, A. (1988), 'Contextual transformation in contemporary science', in A. Jamison (ed.), *Keeping Science Straight: A Critical Look at the Assessment of Science and Technology*, Gothenburg: University of Gothenburg Press, pp. 59–85.

Rip, A. and Nederhof, A. (1985), 'Between *dirigisme* and *laisser-faire*: effects of implementing the science policy priority for biotechnology in the Netherlands', *Research Policy*, **15** (5), 253–68.

Rossiter, M. (1993), 'The (Matthew) Matilda effect in science', *Social Studies of Science*, **23**, 325–41.

Shinn, T. (1988), 'Hiérarchies des chercheurs et formes des recherches', *Actes de la Recherche en Sciences Sociales*, **74** (sept) 2–22.

Siow, A. (1998), 'Tenure and other unusual personnel practices in academia', *Journal of Law, Economics and Organization*, **14**, 152–73.

Wagner, I. (2006), 'Coupling careers in intellectual and artistic worlds', *Qualitative Sociology Review*, **2** (3), 78–98 (online).

Wiener, N. (1956), *I am a Mathematician. The Later Life of a Prodigy*, New York: Doubleday, p. 87.

Woolgar, S. (1976), 'The identification and definition of scientific collectivities', in G. Lemaine, R. McLeod, M. Mulkay and P. Weingart (eds), *Perspectives on the Emergence of Scientific Disciplines*, The Hague and Paris: Mouton, pp. 223–45.

Xie, Y. and Shauman, K. (2003), *Women in Science: Career Processes and Outcomes*, Cambridge, MA: Harvard University Press.

Zuckerman H. (1968), 'Patterns of name-ordering among authors of scientific papers: a study of social symbolism and its ambiguity', *American Journal of Sociology*, **74**, 275–91.

Zuckerman, H., Cole, J. and Bruer, J. (1991), *The Outer Circle: Women in the Scientific Community*, New York: Norton.

6 Society's influence on knowledge content

In the previous chapters, we explored sociological approaches to the phenomenon of science that explain social structures and dynamics, but fail to address scientific content. In this chapter, we shall examine a number of works concerned with identifying the extent to which the knowledge system and scientific content (data, concepts, theories, methods and so on) can be explained by social factors. The first two sections broach the issue from an overall standpoint. The third section presents intermediate sociological analyses focused on the emergence of scientific specialities and the recurrence of problems in a particular research field. The fourth to sixth sections look at social studies of scientific knowledge (Sociology of Scientific Knowledge: SSK) that were influenced by relativist theories: conceptual foundations, programmes, critiques and extensions. Finally, the seventh section demonstrates how the research community that specialises in the sociology of the sciences has changed its approach to these questions.

The Macrosociological Science Trend

A first means of addressing scientific content is by analysing how the knowledge system's key trends have evolved and how scientists' areas of interest have changed. This is a macroscopic analysis. The major trends and the fields in which knowledge develops can be explained partly in sociological terms. However, the actual breakdown of knowledge is better explained through its nature, logic and methodology.

Society's Influence on Knowledge Systems

Many authors have suggested that society influences the nature and content of scientific knowledge. Thus, Condorcet, Comte and Marx established a relationship between the structures of society and those of knowledge systems. The scientific mind is dependent on the state of society. According to Auguste Comte, science is linked to a specific organisational form of society in which work is organised (in factories) so as to maximise productivity rather than to respect a custom. Karl Marx, on the other hand, believed that modern science was born from the demands of capitalism and its production methods. The use of machines and the gradual improvement of production and its efficiency can only be correlated if knowledge falls into line with the development of know-how.

Based on ethnological data, Émile Durkheim (1858–1917) reflected on

the social conditions for the emergence of categories and forms of classification. He showed that this occurs as a result of social experience. Humans class
living beings according to the distribution of clans in society and the struggles
between social groups. Moreover, primitive forms of classification influenced
the first scientific classifications. Scheler (1926) also underlined the sociological
nature of all forms of thinking and knowledge, these being dependent on the
type of working and trading community in question, with a patriarchal community being at the root of science. The nature of knowledge, its acquisition
and preservation mechanisms and its role in society all depend on its economic
relations.

Based on their analysis of twentieth-century society, Nikolai Bukharin (a
philosopher) and Boris Hessen (a physicist and historian) (1931), influenced by
Marx and Lenin, concluded that science, ideology and economic infrastructure
were interlinked. According to them, science is geared towards serving military
and capitalist industrial interests. Scientific methods and reductionism are inspired
by bourgeois ideology and reflect the interests of this social class. Research
methods are determined by the tasks to which the bourgeoisie grants priority. The
principle of 'pure science' is simply an ideology. These authors inspired a group
of British humanist scientists (Bernal, Joseph Needham and the Social Relations
in Science movement) to develop a systematic sociological analysis of science. In
The Social Function of Science, Bernal (1939) put forward the idea of planning
science in a way compatible with a non-capitalist economy. This work prompted
strong reactions within the scientific community and led the physicist Polanyi to
found the Society for Freedom in Science, which opposed any social channelling
of scientific research and developed the idea of scientific autonomy (autonomous
government).

Other thinkers have attributed varying degrees of importance to the influence of society, according to the social group. Lukes (1973) suggested that objective knowledge can only be achieved from certain positions in the social system,
for example, within the rising classes, whose struggle for emancipation does not
require them to distort reality. Florian Znaniecki (1940) demonstrated the existence of a relationship of functional dependence between the social roles played by
scientists and the type of knowledge they produce.

The 1930s also witnessed a proliferation of attempts to explain the content
of scientific theories in terms of social conditions (class interests, racial origin
and political ideology). These explanations vanished after the Second World
War, giving way to the general conviction that the scientific problems taken on by
scientists are determined by the resources available and the organisation of work
within scientific communities, rather than external influences. According to Ben-
David (1991), these influences are restricted by internal dynamics. The scientific
historians Elkana (1968; Mendelsohn and Elkana, 1981) and Koyré (1958) highlighted the indirect effects of external conditions, but avoided making a sociological generalisation. Kuhn (1962), however, suggested that the influence of external
conditions is stronger during periods of scientific crisis, when the fertility of a
scientific paradigm begins to wane.

Influence on the Choice of Research Topics

According to Merton and Sorokin, the topics on which scientists focus their attention are influenced by society's current preoccupations, due to the interdependence between science and its social environment (the dominant culture and the value system). These influences are all-encompassing, as they govern the major trends of the entire knowledge system.

Sorokin (1957) believed there to be three cultural systems: spiritualism (reality is beyond the senses), sensualism (reality can be reduced to what is perceived by the senses) and idealism (reality is a combination of perceptible and imperceptible elements). His contention was that these systems condition the scientific problems addressed and have fluctuated over the course of time. Throughout history, science has focused alternately on spiritual and material phenomena. The rise and fall of interest in research on light and the rise/fall in the credibility of the different atomist theories over the centuries can be explained by each era's dominant truth system. The attention paid to a subject or theory fluctuates according to the societal context.

The study by Merton (1938) on eighteenth-century Great Britain supports this view. He demonstrated an affinity between puritanical values, the scientific approach and the ethos of science. Research fields and topics are chosen neither at random nor according to intrinsic scientific interests alone, but according to the interests of society. Disparate social and economic influences cause the attention of scientists to converge towards certain topics. This societal influence is multifaceted: direct requests to work on certain problems (industrial – mines, textiles, transport – and military); differential visibility of research work in society; changes in the resources available; social and cognitive densification of certain research fields. Thus, science simply responds to society's demands, as exemplified by the development of astronomy (sixteenth and seventeenth centuries) to cater for navigational requirements, or that of high energy physics in the context of the Cold War of the twentieth century.

Although they operate within a framework that is influenced by the social context, scientists also pursue their own objectives and values. Not all scientific activity is conditioned by society. For example, Merton estimated that 41 per cent of the Royal Society's research efforts in the eighteenth century went to pure research. Ben-David believed that society's influence on the volume of scientific activity is nothing more than marginal, and that the practical ends have little impact on concepts and theories. In his opinion, scientific innovation is linked more to the autonomy granted to scientists than to problem-solving demands.

Is there Continuity or Discontinuity between Forms of Knowledge?

While various authors have demonstrated society's influence on the knowledge system's major trends, they stop short of claiming that knowledge is determined by social factors. The relationship is indirect and the influence partial. However,

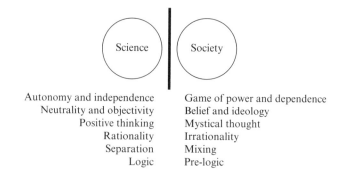

Figure 6.1   *The Great Divide sharing hypothesis*

in the 1970s, a new generation of sociologists argued that the relationship was both more direct and more influential. It was their contention that knowledge is a set of beliefs shared by a social group, the fruit of the scientific output of social constructs, themselves governed by the structure of society (Barnes, 1974). Hence, there is no radical difference between different forms of thinking (the continuity hypothesis): scientific knowledge systems resemble any other system.

The Great Divide Hypothesis

Conversely, other authors considered science to be a distinct form of thinking, and scientific knowledge to be radically different from other types. In the opinion of Comte, the shift from a state of metaphysical knowledge to a state of positive knowledge requires a revolution in social and intellectual terms. Lévy-Bruhl (1857–1939) believed the human spirit to be a stranger to universality. The categories of thought, its functions and methods of reasoning vary from one society to another. The difference between 'positive and rational thinking' and 'mystic and pre-logical thinking that does not follow the non-contradiction principle' is radical. Archaic thinking combines the natural and the supernatural, matter and mind, technology and magic, while modern thinking separates them and respects the principles of logic. These two methods of thinking have nothing in common. Wild thought is irrational and socially determined, while science is rational and universal (Figure 6.1).

 Karl Mannheim (1893–1947) compared the sociology of knowledge to the sociology of ideology (Mannheim, 1967). He distinguished two types of ideology:

- **Limited** ideology relates to a class interest and a partisan imaginary framework.

- **General** ideology relates to mental structure and the possibility of knowing, independent of any class ideology. It relies on a particular social context and on conditions that only make true knowledge possible for an observer who has taken a step back by disengaging from this context. True knowledge requires a social environment that is disconnected from the class struggle: a free-floating intelligentsia that belongs to no social layer in particular, but which aims to distance itself from social concerns, utopias and axiological contaminations (that is, linked to engagements in action).

However, according to Mannheim (1967), the exact sciences are the exception to the rule as they develop according to immanent laws governed by the nature of things and logical possibility. They are free from social determinations and their analysis.

Nevertheless, the question of the 'superiority' of Western scientific thinking has continued to fuel debate, as illustrated by the controversy surrounding the work of Robin Horton (1973) on its differences and similarities with African traditional religious thinking. He argued for the existence of a basis for the universality of reason. According to relativist sociology, on the other hand, these systems of thought are so different as to be *incomparable*, making it impossible to say whether one is truer than the other. The truth is dependent on a system of thought and its social group.

Where Does the Demarcation Lie?

In philosophy, some authors see science as being a space in which knowledge can be objective and detached from all social interests and processes. They answer only to the laws of nature (naturalism), logic (logicism) or experimentation (empiricism and inductionism). Their scientific nature resides in the logical rigour of the observation and verification methods they use (verificationism). Science establishes rational representations of reality, in the form of laws, models and theories. Their validity is independent of the society that produces them. Society can only facilitate or delay their discovery, as evidence for the truth inevitably emerges sooner or later. They lead to a consensus, whereas ideologies have to be imposed. Such 'positivist philosophy' rises against moral and intellectual relativism (in particular that of the anarchist philosopher Paul Feyerabend (1975) and certain sociologists of science).

The thinking of Koyré (1958), Bachelard (2002) and Popper differs significantly from this philosophy in a number of respects. Alexis Koyré (1892–1964) believed that scientific laws are shaped by thought (Figure 6.2), rather than simply reflecting nature. The structures and categories of human thought (linked to biology, rather than philosophy) leave their imprint on observations and laws.

In the opinion of Gaston Bachelard (1884–1962), society influences the origin of knowledge, but science has the ability to break away from both these social contingencies and its own history. The '*travailleur de la preuve*' (as Bachelard describes an individual whose work is based on the quest for proof), nourished by a singular imaginary framework (personal, religious, social and

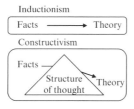

Figure 6.2   *Inductionism versus constructivism*

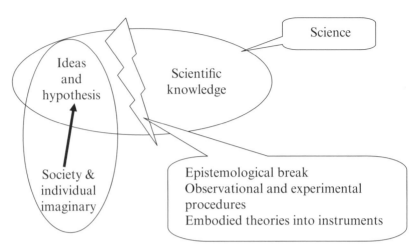

Figure 6.3 *Epistemological break*

artistic) is forced to detach himself from the latter through experimentation and objectivation. Objectivation stems from the establishment of a critical approach that eliminates any epistemological obstacles that prevent scientific knowledge and thinking. An epistemological break is necessary if there is to be a shift from pre-scientific to scientific thinking (Figure 6.3). Such distancing also relies upon the instruments that embody the scientific theories the user must take into account.

Karl Popper (1902–1994) (extending the work of Hans Reichenbach) distinguished between two types of activity: the conception of new ideas and hypotheses (the context of discovery) and their critical analysis (the context of justification,[1] where hypotheses are tested). Knowledge may be grounded in society, but it becomes detached from this grounding through logical reasoning and the testing of hypothetical statements. Testing them against other hypotheses (in other words, comparing the scope of the respective empirical bases explained by these hypotheses) makes it possible to classify them. The influence of social processes is therefore confined to the origin of hypotheses, and fails to impact their scientific validation:

> The question how it happens that a new idea occurs to man . . . may be of great interest to empirical psychology; but it is irrelevant to the logical analysis of scientific knowledge. This latter is concerned not with *questions of fact*, but only with questions of *justification or validity*. (Popper, 1972, p. 31)

Thus, philosophers have suggested the existence of a demarcation (Bachelard's 'epistemological break', Popper's 'demarcation criterion') between science and society. It passes through the realm of scientific activity and marks out a hard core that escapes the eye of society and sociologists alike (Figure 6.4). According to Latour and Woolgar (1979), however, the justification process also relies on contingent conditions and on the dynamics that characterise the revision of hypotheses. This demarcation may not be as clear-cut as it appears.

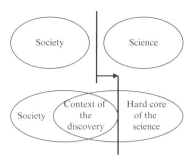

Figure 6.4   *Boundary displacement*

Does Sociological Analysis Have a Role to Play?

Sociology of the sciences has departed from the hard core of science in order to look at what surrounds it (institutions and relationships). It concerns itself with elements of content such as errors, delays, partisan content (for example, eugenism) or deviances (the pseudo-sciences, astrology, parapsychology and so on). Its investigations are confined to institutional frameworks and the study of deviances.

The hard core is beyond the reach of sociologists
It has been argued, on the other hand, that 'true science' is beyond the reach of sociological analysis; it is up to epistemologists to analyse the internal dynamics of science, free from all social influence. Thus, the philosopher Irme Lakatos (1978) called for a project to 'rationally reconstruct the history of science' so as to rid it of its accidents (errors and fallacies) and hence provide epistemological analysis with clean material to work from. In the meantime, sociologists, from Mannhein to Boudon, defended the idea that the validity of certain propositions is objective and independent of social context. Once the obstacles have been removed, the internal logic of development comes into play. If a fertile research programme emerges, it is inevitably adopted, even if scientists are slow to make the choice. From this perspective, sociology confines itself to studying the reasons why certain groups have adopted the right programme, while others have not. It studies the factors that lead to delayed adoption of the right research programme, which would eventually have stamped its mark anyway.

The 'hyper-sociologisation' of science
The above conception restricts the scope of sociological investigations, something a new generation of scientists contested in the 1970s. They launched a campaign against previous analyses (Mertonian sociology and epistemology), and demonstrated the concurrence of social and cognitive factors in knowledge production.

Some rejected the idea of objectivity, claiming that scientific knowledge is purely a social construct. This resulted in the hard core of science being absorbed by sociological analysis, which became the science of sciences. This 'new sociology of the sciences' spelt out its incompatibility with previous approaches by stating that the research process is either relative (strong programme) or contingent (constructivist programme). The ambitions of this form of sociology provoked strong reaction from scientists, epistemologists (whose programme of study on the rational determinants of science would be rendered meaningless) and scientific sociologists influenced by Merton.

However, the new sociology of the sciences did not form a coherent whole. On the contrary, opinions diverged greatly, making it the subject of intense debate. In the 1980s, Latour denounced the inconsistency of sociological relativism, while the PAREX[2] group argued in favour of applying sociological analysis while taking into account science's internal determinations.

The Dynamics of Specialities and Scientific Projects

Instead of addressing the question of society's influence from an overall perspective, empirical research on science has focused on intermediate dynamics, in particular the emergence of specialities and the updating of research problematics. Its aim is to examine the scientific choices made by individual groups of scientists, and to determine the social and cognitive dynamics at play.

The Birth and Development of Scientific Specialities

In Chapter 1, we looked at the emergence of the disciplinary model and, in Chapter 3, the sociological models that explain the birth of new disciplines. The case in point was the concurrence of institutional changes, resource availability and the creation of new social roles, as well as scientific alliances and migrations.

Cole and Zuckerman (1975) suggested that 'cognitive contents' should be taken into account, thus supporting the argument that a speciality is established according to the structure of the academic world and by whether or not scientists perceive there to be a real challenge. Edge and Mulkay (1976) demonstrated that the field of radio astronomy was formed through the recruitment by university physics departments of scientists who had worked for the radar development units of the Bell Telephone Company, in which radio emissions from the Milky Way had been discovered. Interest both in radio physics for industrial or military purposes and in the discoveries made within these units led to the idea that the analysis of astrophysical emissions was a valuable research front. The concurrence of industrial and military investment, new discoveries, scientific migration and the scientific community's recognition of the importance of this new research front explains the emergence of the field. Specialities result from a combination of resources (material, human and cognitive) and pre-existing factors in the environment on which they depend. Cases of simultaneous discoveries have led to the belief that, at a given moment in time, the same factors can be at play in several locations. Thus, new discoveries may be conditioned less by nature than by the concurrence of several factors: the availability of an instrument, access to samples, cognitive resources, the interest generated by a topic, recurring problems encountered in society and the availability of scientific skills.

Conceptual or methodological transfers are also at the root of specialities such as protein crystallography (Law, 1976), which emerged from collaboration between crystallographers and protein chemists. These groups were confronted with the limitations of their speciality (unavailability of equipment, conceptual obstacles, a vein of research drying up and so on) and needed to open the door to other specialities. Scientific migrations due to socioeconomic factors (the geographical spread of teachers around the country in the case of IT development in France) (Grossetti and Mounier-Kuhn, 1995) or internal dynamics (a high number of scientists or the exhaustion of a topic) encourage such collaboration.

The development of scientific specialities relies on their capacity to attract young scientists. This capacity is dependent on contents (themes, challenges) and work methods, but also on institutional, organisational (amount of time spent teaching, access to research resources) and technico-economic (financial and instrumental resources) structures. The Phage Group (dispersed) changed when it started training young researchers and organising seminars and became involved in university teaching programmes (Mullins, 1972).

There is no single training model for new specialities. In their analysis of the burgeoning field of radio astronomy, Edge and Mulkay (1976) observed 'internal ramification processes', whereby scientists identify new problems and multiply their lines of research. In fact, two groups confronted with the same problems act differently depending on their research strategy (eclectic/monothematic) and the amount of technical investment received, but also on the degree of interdependence and competition between the groups. They differentiate themselves in order to reduce the pressure of competition.

The Redirection of Research Projects

To examine the series of choices and decisions made by scientists in terms of research topics and the methodological approaches implemented, it is important to come into contact with them, follow them as they go about their activities or build highly detailed descriptions of them.

Contingency, external influences and the protective role of laboratories

Analyses of this kind highlight the importance of local internal and external contingencies,[3] notably accident and chance (Barber and Fox, 1958). Reference has been made to the case of a scientist who lacked sufficient quantities of the enzyme he was studying, but decided to pursue his work using another enzyme that came to hand. Incidentally, he observed that the ears of the rabbits he used in his experiments softened, before returning to their usual stiffness. Nevertheless, he continued with his original project. Seven years later, during a discussion with a colleague, he remembered the incident and went on to ask his students to think about the problem.

This type of factor, which clearly affects the course of research, is erased from the reports published. Barber and Fox (1958) talked about the 'retrospective falsification' that explains the differences between the way research is actually performed and the way it is presented in publications. Thus, Feltz (1991) reported on the case of a biologist who, when faced with difficulties in acquiring sheep foetuses, decided to use hamster foetuses, considered to constitute a good alternative model. This change prompted the scientist to alter both her line of questioning and her work methods, but also to produce results of another type. Confronted with a lack of resources, she reorganised her work and the structure of the problem she wanted to address. In her thesis and publications, however, the story was reconstructed. It may have been less faithful to the historical truth, but it stuck more closely to the professional standards that are the basis for logical and consistent scientific research. Many other contingency factors have been identified by sociologists, such as inappropriate initial training, the priorities of a laboratory director, the availability of an instrument and so on (Lemaine et al., 1969).

During the course of a research process, a significant number of random factors are at play, whether they relate to the research itself (non-materialisation of the expected results), local conditions (instrument malfunctions or unavailability, interference with the work of another scientist) or external resources (failings or a change of strategy on the part of an associate, redefinition of funding priorities). Such contingencies threaten the success of projects. In reality, the laboratory puts a mechanism in place to protect and stabilise research projects in the face of unforeseen environmental factors (Vinck, 1992). It helps keep projects afloat in spite of any rearrangements they may suffer and any changes in direction they may be forced to make. When a resource is lacking, the laboratory assists in the mobilisation of substitute resources (for example, by redirecting funding from other projects that have cash to spare, by assigning a student to the topic or by using a model that is already known to the team). When an external actor on which the project depends alters its priorities, the laboratory deploys a strategy to minimise this extrinsic influence. It therefore protects high-risk scientific investigations against economic, social and institutional circumstances, and alleviates the determinant nature of these factors. The laboratory is a device that tempers external influences.

The strategy of researchers: highly cautious redirection

Researchers are also strategists within their own field, as Bourdieu (1975) and Latour and Woolgar (1979 [1986]) theorised. Their strategies are both intellectual (the possibility of solving a problem) and social (depending on the recognition they are able to obtain from their peers) (Lemaine et al., 1969). The level of visibility of research subjects plays an important role in the orientation of their projects. Strategies vary according to the basic circumstances of the research (the hypothesis may be probable or improbable, the paradigm well established or non-existent), but also as a result of personal preferences in terms of risk taking, career security, risk sharing and so on. Some scientists choose to pursue a risky, but promising hypothesis, others work simultaneously on high- and low-risk hypotheses, while others again diversify their activities or fall back on less well-trodden projects. The differentiation of scientific activities is also geared towards the creation of less competitive areas of activity (Lemaine et al., 1969; Edge and Mulkay, 1976).

The diversity of strategies (diversification, substitution, disengagement) is, however, limited by the search for the 'right balance' between conservatism and a radical shift in research problems (Gieryn, 1978). The emphasis on productivity in terms of publication output, the perception by institutions that persistence is a form of excellence, and the fact that recognition in a particular subject area provides access to new resources, serve to explain the limited success achieved by radical redirection strategies. Furthermore, external demands from society, not to mention research institutions, may have a limited influence and struggle to redirect the projects of research teams.

Box 6.1   *Social determination of scientific knowledge*

Ideological and cultural prejudice, social movements and the position scientists occupy in society go some way to explaining scientific constructs. The history of quantum mechanics (Forman, 1980) is fairly instructive regarding these points. In 1919, physicists suddenly distanced themselves from the notion of causality, which is central to physics. They went as far as to repudiate the principle, several years before quantum mechanics was first founded.

As far as the rational, internalist history of science is concerned this shift was due to internal discussions on the concept of causality within the field of physics, initiated by James H. Jeans and Poincaré in 1910–1912. However, the response to these discussions was limited. Forman, on the other hand, introduced sociological factors to the explanation, in particular the influence of the dominant ideology and the value crisis that had hit society. The rejection of the classical conception of causality and the adoption of a new conception (indeterminism) are indicative of the way in which scientists defended themselves against attacks from society.

Military defeat and the signing of the Treaty of Versailles at the end of the First World War left Germany in shock. From a state of unwavering belief in the supremacy of their Empire thanks to science and industry, the Germans were suddenly stripped of their illusions. The values (such as scientific rationalism) that accompanied economic and industrial expansion were called into question and triggered a return to romanticism and spiritual renewal. The notion of Destiny was placed back on the agenda and set against that of causality. A new hostility to science emerged.

Mathematicians and physicists felt targeted and reacted by taking part in the debate. Several of them identified with spiritual renewal and engaged in self-criticism. Some founded associations to defend their scientific status. Others, such as Einstein, argued that scientists should firmly maintain their belief in determinism. But many more publicly rejected and repudiated the notions of causality and determinism.

Scientists had to renegotiate both their role and the recognition offered to them by society, as well as justifying their existence as a socioprofessional group (even with Germany slumped deep in misery at the time). They did so by translating the social value crisis into a crisis within their science. They developed arguments and scientific demonstrations to bring their science up to date. They responded to society's questioning with this 'new science'. Some acknowledged that they had run out of rational arguments against the rise of occultism and esotericism. A number of them made ideological concessions to anti-rationalism and the pervading atmosphere of irrationalism. They adopted vocabulary reminiscent of the cabal and talked of the mystical properties of numbers (with respect to theoretical spectral analysis, a very fashionable topic at the time).

Scientific knowledge itself was also affected. The crisis hit even the best-established theories, including Newtonian mechanics and Euclidean geometry. The social value crisis became a crisis of fundaments. The greatest scientists launched headlong into the complete restructuring of the fundaments of their discipline. In mathematics, this reconstruction paved the way for intuitionism. In physics, some extended the theory of general relativity to the whole of physics. In 1921, several physicists converted to non-causality. Some gave speeches to announce that they had turned their back on the doctrine of causality and that a new era would soon begin, during which physicists would free the world from the shackles of determinism. In 1925, Werner Heisenberg developed matrix mechanics and in 1926, Erwin Schrödinger founded wave mechanics. Heisenberg established the uncertainty principle, while Niels Bohr set out the principle of complementarity. In doing so they created a scientific basis for their renouncement of causality. They then transcribed their discoveries for the public. In fact, Heisenberg published the vulgarised version of his work even before his technical article appeared. In 1928, Bohr declared that there was no room for freedom in previous determinist and mechanical conceptions of matter, be it in the form of free will or a higher power, whereas the new physics attributed altogether different characteristics to the universe. Bohr spoke of the irrationality of physical phenomena, while Ludwig von Bertalanffy wrote that modern physics cleared the way for a new type of mysticism.

It could therefore be said that scientific content was the scientists' answer to the social crisis threatening them. Their research reflected the social pressures placed upon them and presented the basis for a new contract with society that would enable them to justify their status. In this analysis by Forman, the factors explaining the emergence of quantum physics are social ones: the dominant ideology and the movements spreading through society.

Conceptual Bases of the Sociology of Scientific Knowledge

In the 1960s and 1970s, a new generation of sociologists defined science as a multitude of local cultures in which scientists conform to the standards, values and local interests determined by social structures. They laid the foundations for the Sociology of Scientific Knowledge (SSK), which drew its concepts from the work of philosophers such as Wittgenstein, Duhem and Quine, the historian Kuhn and classical sociology. We shall now present a few of these theoretical bases.

The Sciences as Language Games and Forms of Life

Two works by Ludwig Wittgenstein (1922 [1961] and 1953 [2001]) operated a reversal in the doctrine according to which logic is the keystone of the scientific approach (logicism). The latter publication, which drew the attention of sociologists, placed language at the heart of the analysis. It suggested that the meaning of words is not set in stone, but varies according to their use. Meaning is connected to usage. Statements take on meaning from the activities to which they refer. The meaning of '3D' (three-dimensional), for example, varies according to the graphical practices of computer game designers, developers and authors. The definition of a door differs depending on whether it is used by a mason (a hole in a wall, measured in centimetres) or a carpenter (a hinged opening, measured in millimetres).

Each activity corresponds to a different 'language game', such as describing an object according to its physical appearance, reporting on an event, presenting the results of an experiment using tables and diagrams, asking for information or performing a demonstration. Thus, Wittgenstein paved the way for the study of scientific practices and gave a few pointers for their analysis: think about the aim and the circumstances in which something is said, examine the forms of action that accompany these words and the arena in which they are used.

The language game is determined by its usage rules. The novice discovers these by watching how the players play the game and by identifying the family resemblance between different parts of the game. The novice has understood the rules of the game from the moment he/she learns to apply them; their application is a prerequisite for their comprehension. These rules are tacit and local; they depend on the way the game is played. They also evolve in parallel with the gradual transformation of the activity. Thus, sociologists were invited to focus their attention on examining local situations, the way in which scientific research is played and replayed and how the rules of the game are redefined locally. These rules require the renewal of the social agreement governing their modes of application. The sociologist David Bloor (1976) developed a programme to analyse scientific practices (drafting of publications, drawing up of observation reports or application of mathematical formulae) as though they were specific language games whose tacit rules and conventions need to be understood (social consensuses that serve to codify scientific research).

Interpretative Flexibility and How Conventions Condition Facts

Two other philosophers provided sociological analysis with access to scientific content: Wilhem Quine (1974) and Pierre Duhem. By undermining the ideas of simple rationality, the primacy of logic and the irrefutable evidence of facts, they created conceptual points of entry that were then exploited by the sociology of the sciences.

How theories condition facts

Facts in themselves rarely lead to proof of hypothesis or to their refutation. Scientists are wary of the apparent and illusory 'evidence of facts'. Facts are only deemed to be significant or valid once various interpretation, evaluation and qualification processes have taken place and through their connection to prior knowledge. As a result, observation loses its primary role, and is instead assigned to the interpretative framework that allows facts and data to be qualified. Observation is dependent on accepted theories and the sociocognitive factors used to interpret them (accepted conventions, the language used and background knowledge). Fact is also indissociable from the way in which it is expressed (linguistically speaking, in particular), which carries both meaning and interpretative elements. Organising and classifying facts requires there to be a concept in place. The identification and isolation of a phenomenon or object from the flow of sensory perception also implies that the observer has concepts at his/her disposal. Categories of thought therefore make their own imprint on observations. Experimentation is always accompanied by interpretation of the phenomenon. Raw data already constitute an interpretation.

In addition, experimenters carry out adjustments so as to obtain satisfactory data. These corrections play an important role in the production of 'raw' data precisely because they are guided by the interpretation of the phenomenon. Interpretation, far from following on from observation, actually precedes it.

The under-determination of theories by facts

According to Quine (1974), a number of theories may serve to explain the same set of observations. They can be both logically incompatible with one another and empirically equivalent and compatible with the data. A set of facts does not therefore allow one theory in particular to stand head and shoulders above the rest. It can be said that theories are underdetermined by observations. Following in this vein, Duhem (1906) observed a disparity between concrete fact and theoretical fact; between the two, there is indetermination. A single practical fact can be linked to an infinite number of theoretical facts and vice versa. Such ideas prompt sociologists to suggest that if observations alone are not sufficient to determine the theory that best interprets them, then other factors, in particular social ones, must come into play. The existence of several possible interpretations can be a source of disagreement between scientists as to which interpretation is correct. If one interpretation stands out it must be explained according to the discipline's usual rules of interpretation (Duhem), with respect to cognitive or aesthetic factors (the beauty and simplicity of a theory), or even extra-scientific factors. For example, it may relate to a set of decisions taken by the scientists themselves. Thill (1973) observed how high-energy physicists were compelled to lay down a 'pure convention', rooted in practice rather than epistemology, to set the average particle beam contamination value to be taken into account in their research. Because the two methods employed to assess this contamination produced divergent results (despite the fact that the calculations were verified), the result used in their subsequent research was set conventionally.

The circularity of procedures to evaluate theories through experimentation

Experimental tests are always ambiguous. An experiment alone cannot disprove a theory. When it disagrees with the theory, one can only state that it calls into question one of the hypotheses of this theory,[4] a component of the experimental set-up used or the competence of the person performing the experiment. Harry Collins (1974) spoke of the 'experimenter's regression': if the results clash with the theory, it is impossible to say whence the error stems. In practice, scientists do not call their work into question if inconsistencies arise. They trust the conventions of result interpretation and a part of the accepted theories. Certain elements are accepted without question, in particular when they relate to conventions accepted within a specific scientific community.

For theories to be tested they must be held up against the facts, but establishing and observing these facts inevitably leads to their interpretation. It is therefore almost logically impossible to escape the framework of interpretation, which comprises accepted theories and conventions, as well as beliefs relating to reality. Scientific procedures display circular characteristics ('experimental circularity').

The legacy of Duhem and Quine

In the work of Duhem and Quine, the sociology of scientific knowledge found conceptual bases for the sociological analysis of contents. It found arguments that demonstrate that nature and logic alone cannot explain accepted scientific theories. Among Duhem's analyses, sociologists clung on to those of disparity (between concrete facts and theoretical facts) and interpretative flexibility. They concluded that scientific factuality is contingent, that it is an illusion of neutrality constructed by scientists who obscure the details of their concrete actions and the local historicity of fact production (Michael Lynch, 1985). They asserted that the relationship between the muddle of observed facts and the clarity of facts related in publications is also contingent. Duhem defended the idea of a more pliable and contextual rationality. First and foremost, sociologists who studied his work highlighted the role of contingent or conventional factors in explaining the success or failure of a theory.

Conventions and Local Cultures

Experiments are also dependent on the local culture of each scientific group, in particular regarding their way of applying protocols (Fleck, 1935 [1979]). Knowledge, which is built on operative practices, including those that involve symbolic entities, relies on methods, technical skills and tacit know-how (Polanyi, 1958) that experimenters do not always make explicit. Experimental practices are not transparent. Data and their interpretation are produced locally and are affected by habits and reasonable agreements (the conventions in place within a group of scientists) (Kuhn, 1962). Thus, a theory's validation equates to the incorporation of a new element in a set of social conventions that have already been accepted within the group. From that perspective, data and theories are the products of a culture embodied by the conventions of a social group. These conventions form a framework,[5] a sociocognitive structure that gives meaning and consistency to facts and scientific concepts. The framework (constructed from the relationships between background knowledge, conventions and beliefs) makes it possible to differentiate and classify what is perceived and to link this up with other elements. It is an interpretative model without which observations would make no sense.

From the moment 'raw facts' are produced, various logical, methodological, instrumental, cognitive, aesthetic, conventional and cultural factors enter the equation, shaping the facts and their interpretation. An examination of working practices that use observation instruments (microscopes, telescopes or radiography) shows that the images and traces produced are difficult to decipher (Box 6.2). They require learning, experience in the art of observing, rules and conventions to guide the way they are produced and read. Observing does not simply involve letting perception do all the work. On the contrary, researchers must seek what there is to see. They construct the object according to their working habits, the knowledge they possess and the nature of their project. They must know what they are looking for in order to see it. Observation data are linked to the object anticipated.

Box 6.2   *The discovery of N-rays*

At the turn of the nineteenth century, radiation was a new phenomenon that generated considerable interest. Wilheim Röntgem discovered X-rays and Becquerel the radiation emitted by uranium. The subject was an extremely fertile one and enjoyed considerable exposure in society. The search for new types of radiation was at the forefront of science. A conceptual model was already becoming firmly established and was guiding scientists in their way of thinking about the radiation they were seeking and in the processes required to discover it. It conditioned thinking, actions, but also observation. Observers already had an idea of the shape of the object before they came across it. Indeed, nobody was surprised when, in 1903, Blondot discovered N-rays. The discovery was logical, not to mention timely for France, which had been hoping to add another prestigious name to the international scene after those of Becquerel and Pierre and Marie Curie. The discovery was logically, empirically and sociologically probable. The experiment was reproduced. A number of scientists from various disciplines confirmed the existence of the phenomenon and worked on its applications. Blondot, who already enjoyed an excellent reputation, became more widely recognised and was awarded a prize by the Academy of Science.

However, not everyone was able to see N-rays. This is normal, as an educated eye is needed to perceive the variations in the flashes of light they produce. Nevertheless, those who were unable to see the rays later succeeded in imposing their point of view; N-rays simply did not exist. No British or German scientist could see them. The French retorted that this made total sense, since it was not in their interest to see them as they would be forced to acknowledge France's supremacy. An American passing through the city of Nancy secretly sabotaged Blondot's instrument and recounted his trickery in the scientific press, claiming that the scientist had failed to notice: he could still see N-rays. Gradually, the American's scepticism prevailed and N-rays were no longer seen by anyone.

The objectivity of what is observed is also the product of intersubjectiv-ity[6] within a scientific community. It is the result of an agreement within the group and of the array of expectations harboured by the scientist, the research programme they are taking part in and those of society with respect to science.

Often, what there is to see is not obvious. Scientists take extreme precautions in establishing and validating the facts, but still continue to doubt their results. Pickering (1981) illustrated this tentativeness by describing the experimental contingencies, the changes made to the apparatus and the hesitations of a group of physicists who had discovered neutral currents at the CERN. Just before the results were published, they were still hesitating. They had doubts about the valid-ity of the experimental set-up and called into question their certainties as soon as they got wind of contradictory information from their American competitors, refusing to make any cut-and-dried statements. The veracity of a fact cannot be assumed but depends on how consistent the fact is with researchers' expectations and on the agreements between them.

Paradigm and the Paradigmatic Community

The work of Thomas Kuhn (1962) opened a new breach that sociologists were happy to enter. His 'paradigm' concept rendered the connection between science's social, institutional and cognitive dynamics tangible, and made it possible to bring social factors into play in the construction of scientific facts.[7]

Merton saw in Kuhn's work a way of understanding the nature of scientific research and its historical developments, even though he believed his sociologi-cal analyses to be weak. He retained the idea of studying small scientific com-munities. Crane (1969) (in Chapter 2) provided a link between the contributions of Kuhn and Merton by re-examining the notion of the invisible college. The Cole brothers (1973) (in Chapter 2) also confirmed the compatibility of Kuhn's analyses, which looked at intellectual factors, with their own analysis of the social structure of scientific communities.

Countering these ideas, the PAREX group highlighted the radical changes Kuhn had triggered by studying the relationships between ideas and scientific communities. Based on an analysis of 12 laboratories, Lemaine et al. (1972) con-cluded that Kuhn's notion of normal science masked contrasting realities from one laboratory to the next. Nevertheless, they remained prudent *vis-à-vis* the influence of social factors on scientific content.

British sociologists such as Barnes (1974) and Bloor (1976) considered, on the other hand, that Kuhn's contribution seriously undermined the Mertonian approach, as it highlighted the cultural nature of scientific activity. The notion of paradigm is tied to the beliefs and conventions in place within a social group. Barnes (1982) saw strong parallels between his work and anthropological studies, while Bloor highlighted its similarities with Wittgenstein's 'forms of life'. Their attention was therefore focused on cultural variations in the use of scientific categories and theories.

The paradigm or the underlying social influence

'Paradigm' means model. A scientific group's paradigm is the set of common traits shared by its members (models of behaviour, action and thought), and on which they forge their identity. This notion invites us to consider different ways of seeing things (models of thought). The paradigmatic community is a community of perception. The paradigm conditions the birth and development of a scientific field and the development of its theoretical corpus, because it guides scientists in their work. It is a way of viewing the world and organising reality. It structures the way in which science is conducted and leaves a cultural imprint on scientific statements.

Young researchers assimilate the paradigm by understanding how a problem is best posed and studied, and by learning exemplary scientific approaches and the most effective working models, both during their training and when they take their first steps into research. Learning is conditioned by teaching methods, the examples in manuals, didactic experiences, the accounts of exemplary experiments, the comments of colleagues and superiors, as well as anecdotal evidence. The progress of young researchers is linked to their ability to imitate their elders. The paradigm is propagated through illustrations, textbook examples, instruments and reference texts, as well as concepts, basic axioms, theories, judgement criteria and the works of exemplary scientists of the past.

Every speciality is characterised by a paradigmatic (or disciplinary) matrix comprising:

- **Symbolic generalisations**: commonly employed expressions, which are often formalised (such as U = R.I).

- **Metaphysical paradigms**: common beliefs (such as 'All perceptible phenomena are due to interaction between atoms') of an ontological nature – that is, concerned with what is real (for example, time is a measurable quantity) – or of a heuristic nature (they allow interventions on things to be interpreted and conceived). They also include the group's preferred metaphors or analogies, linked to the definition of the object of research (molecule or cell, person or group, phenomenon or statement).

- **Values**: these give individuals a sense of belonging to a group. They specify what qualities a successful result should have (for example, only quantitative results are valid), as well as working rules (which groups and networks to move around in), organisational methods, the working philosophy, acceptable social practices (consultation, the confrontation of reality in the field, vulgarisation), the type of report required, the writing style and the right methods of representation (tables, graphs, images).

- **Paradigmatic examples**: examples of a typical problem and a concrete solution presented to students (in manuals, during practical work or as a subject of analysis). They guide working methods and cover typical problems young researchers must learn to solve, as well as working techniques, observational tricks, methods of reasoning and the language to be used.

The notion of paradigm opened up new ways of analysing science, by providing an understanding of the social and cognitive identity of scientific communities. However, the notion raises a number of problems: (i) the connection between a social group and worldwide view is ambiguous. In some cases the starting point for distinguishing between different scientific groups is the paradigm while, in others, the search starts with a specific group and tracks down the paradigm from there; and (ii) the term 'paradigm' has approximately 30 different meanings in Kuhn's publication.[8]

The incommensurability of paradigms
Paradigms are closed in on themselves. They have their own evaluation criteria. Theories, data and procedures can be neither understood nor criticised based on another paradigm. Paradigms can be neither interchanged nor held up against each other, as there are no universal comparison criteria. Nor can there be a crucial experiment that allows paradigms to be compared, as experiments are necessarily designed and interpreted within a particular paradigm.

Every paradigmatic community is isolated in terms of information (set of problems), norms (problem resolution standards), semantics (the meaning of concepts is linked to the relationships within the paradigm) and ontology (worlds/worldviews are specific). The incommensurability between these belief systems also has a bearing on the relationship between science and religion. The elements of one system are meaningless to the other, making it impossible to say whether one is truer than the other. The notions of truth, proof, logical criteria and reason lose their status as universal points of reference. Feyerabend (1975) concluded that it is impossible to express the concepts of a theory using the terms of another. No argument can justify favouring science over other knowledge systems.[9]

Normal science
The sciences experience normal phases during which scientists work within the paradigm defining the enigmas to be solved, such as:

- **Producing significant scientific facts**, for example: calculating a planet's position, determining the atomic charge of a new element, producing an element with the characteristics specified in Mendeleyev's table, establishing the spectral form of a compound, identifying the *habitus* of a social group. When Guillemin and Schally joined the field that would later earn them the Nobel Prize, the paradigm was already established: the hypothalamus produces hormone-releasing factors that control the pituitary gland. All that remained was to define the chemical nature of these substances, and how to isolate, purify and analyse them.

- **Demonstrating that the facts are consistent with the theory and building instruments to do so**: for example, the telescope, to demonstrate the stellar parallax predicted by Copernicus, the particle accelerator, to produce the elementary particles described by the theory, and large scintillation counters, to demonstrate the existence of neutrinos.

- **Constructing a theoretical model**, again within the paradigm, is another alternative making it possible to report on certain observations. In the example of Guillemin and Schally, the enigma involves showing that the substances isolated and analysed are consistent with those set forth in the paradigm, in addition to synthesising analogous factors and analysing the action mechanisms. Having solved this enigma using a particular factor (the TRF), similar enigmas using other factors still had to be solved.

- **Improving theories**, for example: improving the precision of the Planck constant or the Avogadro number, establishing quantitative laws linking together several variables defined by the paradigm or finding a similar but clearer mathematical formulation. In the example of Guillemin and Schally, having resolved the previous enigmas, they still had to describe the relationship between the hypothalamic factors, the hypothalamus and the pituitary gland.

Thus, failure can be explained by the scientist's inability to resolve the enigma, rather than by the inadequacy of the paradigm, which is accepted and undisputed. The paradigm forms the framework of a research tradition from which the scientist does not waver.

Paradigmatic change

Scientific disciplines experience both phases of continuous progress (normal science) and revolutions. As anomalies and inconsistencies accumulate, they enter periods of crisis during which the paradigm is called into question. Scientists rise up against the authority of their tradition. This leads to a frenzy of intellectual activity as the community searches for new paradigms.

Mullins (1972) established a relationship between the state of the social networks of a scientific field and its pragmatic evolution. He distinguished three phases:

- The **paradigmatic phase**: a group is formed comprising a few renowned scientists working independently. They do not stand out from their social milieu as a group and there is no formal communication between them (loose group).

- The **dogmatic phase** (communication network): the members of the paradigmatic group meet, exchange information (apprenticeship), collaborate (colleagueship), publish jointly (co-authorship) and quote each other. Their range of problematics is narrowed. At the *network stage* (loose links, informal exchanges regarding methodological trends), the exchanges and consensuses they produce are determinative because they define tasks, validity criteria and the required resources (technical and human). They establish the barriers to entry into the network. At the *cluster stage*, the researchers establish common standards (research language and protocol, signature rules, management procedures). Their time is taken up by the new activity, which commands significant resources. They sometimes find themselves in competition.

- The **academic phase**: more researchers join the field, inspired by the initial success achieved and encouraged by the stabilisation of procedures, techniques, languages and hypotheses. Competition becomes fiercer as the field becomes structured and institutionalised, with the setting up of conferences, journals, university courses and reference manuals. The field receives regular structural support. Nevertheless, other researchers continue to believe in the virtues of the previous paradigm. The new paradigm triumphs not only because it succeeds in convincing its detractors, but because these eventually die out.

The idea of a paradigmatic revolution calls into question the idea of cognitive continuity. The switch from one paradigm to another is, in essence, an irrational 'mystical conversion' (indeed, Lemaine criticised Kuhn for his Gestalt conception of paradigm change), which can be explained by extra-scientific factors. Popper asserted that a theory is abandoned in favour of another if the empirical basis for the facts explained by the new theory is stronger than that of its predecessor (the degree to which theories can be backed up empirically). As paradigms are incommensurable, this critical conception sees its foundations crumble. Furthermore, rather than the idea of science progressing continuously, the vision of science conjured up by Kuhn's analysis is one made up of normative traditions; when they approach a problem, scientists do not disregard everything that has happened in the past. On the contrary, the institutionalised paradigm standardises their work. To be recognised, they must prove themselves within this framework. The paradigm is of a normative nature both from a social point of view and with respect to content.

The Principles of Relativist Sociology

Based on the conceptual foundations presented above, David Bloor (Strong Programme) and Harry Collins (Empirical Programme of Relativism) formalised a new programme for the sociological analysis of the sciences.

The Strong Programme and the Symmetry Principle (Edinburgh)

The relativist movement (Box 6.3) described sciences as belief systems relative to the social groups subscribing to them. From this perspective, the convictions of nuclear physicists are no less sociological than those of African witch doctors. Knowledge is a conventional belief. It can be explained by the social groups that construct it and by the social interests driving them.

The relativist movement's programme of research, which was borne out of Edinburgh University's Science Studies Unit (Barry Barnes, David Bloor and David Edge, followed by Donald MacKenzie, Pickering and Steve Shapin) was formalised by Bloor in 1976 and qualified as a strong programme for the sociology of the sciences, its aim being to remove the inhibitions of sociologists with regard to natural

---

Box 6.3   *Fundamental propositions of the relativist analysis (Mulkay, 1980)*

1 The meaning of a fact (observation, calculation) is dependent on the hypotheses shared by the group.

2 The acceptability and acknowledgement of a piece of knowledge depends on the social context.

3 The repositories of meaning marshalled by scientists are drawn from the social context.

4 The rules of reasoning depend on informal social negotiations.

---

sciences. It has been applied to reconstructed situations through socio-historical analyses (Robert Boyle's vacuum pump as studied by Shapin, the statistical controversy between George Udny Yule and Karl Pearson as studied by MacKenzie, phrenology and so on) and contemporary situations (gravitational waves, quarks, parapsychology and so on). These have led to scientific controversies that have provided access to arguments and facts, as a result of being debated publicly. The sociologist describes the conceptual systems and practices of the scientists in question, and resituates them in their macro-social, political, religious and economic context. Thus, scientific constructions can be explained based on the social context.

The principles of the strong programme
The strong programme extends the supposed epistemological principles used in other sciences to the sociology of the sciences (naturalist approach to the production of knowledge):

• The **causality principle**: this involves determining all types of conditions and causes that may explain the emergence and development of knowledge.

• The **impartiality principle**: the scientist must avoid prejudging whether or not a piece of knowledge is true, or whether a belief is rational or irrational.

• The **symmetry principle**: the sociologist must look for the same types of cause to explain true beliefs and false beliefs.

• The **reflexivity principle**: the explanatory models used to report on sciences must also apply to the statements arising from the sociology of the sciences.

Often, these principles are only partially applied. Moreover, the strong programme has been the subject of numerous debates both within and outside the Edinburgh School. The symmetry principle still provides fuel for intense debate.

The symmetry principle
The symmetry principle opposes the blatant asymmetries found in certain analyses of science, in which the authors rely upon the rationality, elegance and clarity

of a demonstration to judge its success. Others underline genius, rigour and the fact that scientists have broken free of prejudice, irrationality and social influences and have grown attentive to nature, unlike those who have been misled. Some authors distinguish between Isaac Newton's genius, a model of rationality in the field of astronomy, and his irrationality, steeped in the mystical and in astrology. Such analyses are asymmetric; they resort to rational arguments to explain the success of a discovery and to an analysis of social influences to understand any deviances. Indeed, Martin Hollis (Hollis and Lukes, 1982; Hollis, 1988) claimed that rational knowledge requires one kind of explanation, while false and irrational belief requires another. Those theories that are discarded can be explained based on the theories that have superseded them. What is more, these new theories harbour the truth that allows the errors of their predecessors to be explained. There is little purpose in studying the knowledge production process; only a posteriori analyses are relevant.

Such asymmetrical explanations are unacceptable according to Bloor (1976). The symmetry principle requires that we use the same causes to analyse both knowledge that is accepted and that which is rejected. It is not acceptable to explain scientific theories through empirical data drawn from nature, method and logical reasoning, while explaining erroneous theories based on psychological and social factors. Social factors come into play in both cases. The aim should simply be to report on these.

The symmetry principle (Box 6.4) is a rule of method. It does not postulate

---

Box 6.4   *A few versions of the symmetry principle*

- Application restricted to analysing social factors alone. Socio-epistemologists accept a more flexible version of the symmetry principle that shows how social factors might intervene, but they also consider that cognitive factors make the ultimate difference.

- Explanation of true and false beliefs based on social factors. The aim, for the strong programme, is to remain agnostic and to refuse to believe in the existence of cognitive elements determining differences. The protagonists use these cognitive components to support their assertion as to who is rational and who is mistaken. To rely on these components to help distinguish between them would be to adopt a partial point of view. The aim, on the contrary, should be to show that knowledge is a social construct that can be explained by social factors.

- Extension of the symmetry principle to the analysis of the different factors, including those of a cognitive nature. In the opinion of Latour (1987), the observer is invited to follow the examinations, arguments and counterarguments. If cognitive factors enter the debate, one must demonstrate how they intervene and to what extent they influence the outcome of controversies. The final asymmetry can be explained by the accumulation of small differences and by reconstructing all the examinations that lead to a big difference. The symmetry principle involves treating all the elements contained in the explanation equally.

that knowledge and belief are of equal relevance (relativism), but it does require the same instrument to examine both. One must distance oneself from one's cultural presumptions, according to which a fundamental difference exists between true knowledge and false knowledge. The symmetry principle does not exclude the highlighting of differences.

Beliefs and scientific black boxes

In the strong programme, knowledge is treated as a range of beliefs linked to conventional working methods, protocols representing a consensus, conventions implemented through instruments and an established set of habits. The validity of a new piece of knowledge depends on its incorporation in accepted conventions. If peers call into question neither the new construct nor the prior conventions to which it refers, then this new construct will be recognised as a new, accepted piece of knowledge. Its robustness depends on its relationship with previous knowledge. Should anyone cast doubt on the result, they must call into question the protocol, the experimental path or the instruments. If they have doubts about the instrument, they must examine the theory upon which it was constructed. If they have doubts about the theory, they must challenge a series of publications, colleagues and other experimental results that have become widely accepted. All of these elements are social constructs that have become stable. They are the fruit of the work performed by other social groups and of the tacit agreements between them. These social constructs have become obscured because the traces of their construction have been erased. They are 'black boxes' (Latour, 1987), whose boundaries – for example, between fact and opinion, content and context – protect their contents from being called into question (by colleagues and non-scientists who may contest their scientific basis, robustness or legitimacy), on the one hand, and from being relativised (that is, reduced to their local social causality), on the other.

Blondot's N-rays were some way down the path to becoming a black box. Waves were a recognised phenomenon, the instruments were widely known and accepted, the results had been examined and confirmed by others, Blondot was a popular figure and France had been longing for such a discovery. However, before it could be closed, the box became unhinged, the instrument was altered and called into question, the scientist's rigour was brought into doubt and the factuality of the proof was denounced.

The way in which the boundaries are positioned is the result of a process of construction (Gieryn, 1995) and negotiation. The protagonists attempt to impose and legitimise certain distinctions. The observer therefore has a duty to remain agnostic and report on the final construction.

Cognitive interests and professional investment

The Edinburgh sociologists considered that a conventional belief system can only be understood if it is linked to the social interests of the social group in question. A confrontation between theories is essentially a confrontation between groups whose interests (cognitive and instrumental, professional and social) diverge.

These groups are the result of the scientific world's internal structure, for example, the structure linking theoreticians and experimenters, or different specialities. Their cognitive interests are linked to the social investment they have made in acquiring skills and the ability to analyse (Box 6.5): long socialisation process, assimilation of the group's conventions, learning to resolve enigmas within the paradigm. Thus, researchers tend to employ, deploy and defend their cognitive approach.

Box 6.5   *The controversy between orthodox taxonomists and experimentalists*

The social investment made by scientists from both groups explains why each of them challenged other paradigms and defended their own. The more orthodox established their taxonomy based on observation and on a meticulous description of plant morphology. This work required a lengthy learning process both out in the field and in herbariums. It would have been unthinkable for investment on this scale to be made only for a taxonomy requiring a completely different grounding. It was in their *professional interests* to defend their taxonomy, especially considering that it had become operational both at research level and in the cognitive management of the plant world. Professional interests had been joined by *cognitive interests*.

The social and professional investment made by experimentalists was just as considerable (mastering the techniques of biochemistry, molecular biology and genetics) and their taxonomy gave rise to an effective way of managing the plant world. The position of the two groups within the controversy (as regards defining the notion of plant species and the classification of certain species) could therefore be explained by their respective cognitive and professional interests and investment.

Training, gaining experience, establishing social relationships, as well as the creation of instruments and an organisation, are examples of professional investment. They encourage scientists to favour a definition of reality (for example, that of elementary particles[10]) that is most likely to provide fertile ground for the future of their activities. Moreover, when two groups discuss how best to define reality, the controversies that emerge relate as much to their professional interests as to the definition of phenomena. The definition of nature becomes a social stake for these individuals (Dean, 1979).

The commitment of scientists in the eyes of the public is another example of a professional interest. In the controversy surrounding the respective French and American calculations of Neptune's trajectory, it would indeed have been in the interests of the French to highlight the identity of the calculations they used as this confirmed the accuracy of their science, a science whose merits they had just publicly extolled. Similarly, the status of the profession (developing credibility and social recognition) and the dividing line between scientists and the profane (the clergy and amateurs) have been the object of investments that go some way to explaining certain scientific controversies, such as parapsychology (Collins and Pinch, 1993).

Social interests

Knowledge is therefore a resource that serves social groups, the interests of which can be explained by their position in the structure of society (Box 6.6). Thus, knowledge content can be explained by the social position of the groups that produce it and their cognitive and professional interests. The existence of competition between these groups explains the structure of conceptual networks and the fabric of knowledge.

Box 6.6   *The controversy between the statisticians Pearson and Yule surrounding the analysis of the relationship between nominal variables (MacKenzie, 1981)*

Pearson had recently devised a method to study the correlation between two distributions; he suggested extending his method to nominal variables. At the time, Yule was also seeking a solution to the problem of nominal variables. At first glance, their cognitive interests were the same: to develop a method of statistical analysis geared towards nominal variables. However, they argued for 10 years. Pearson wanted to extend his method and establish a unitary theory; Yule rejected this method outright and continued to seek an original solution. Their cognitive interests diverged.

Pearson was actually developing tools that would allow him to establish a working heredity theory (by defining the criteria required for the growth of a healthy population). Yule, on the other hand, was studying correlations with the aim of devising curative instruments. His variables were nominal: life/death, alcoholism/non-alcoholism. He was searching for the tools that were the least sophisticated, but the most suited to the situations he wished to analyse.

Pearson was interested in the theories of heredity because of his ties to the eugenicist movement (improving the human race by preventing the unfit from multiplying). His social commitment and cognitive interests were linked to his membership of the rising social classes, which espoused technocratic ideology and the eugenicist movement that had risen within the new professional classes. Yule, who came from a class of old elite that was beginning to fade, was opposed to these scientific ideologies. Their conflicting cognitive interests could therefore be explained by divergences in their social class interests.

As squabbles crystallised around various scientific questions, the protagonists lost sight of differences between their social aims. They believed the debate was purely scientific. In the controversy surrounding phrenology (Shapin, 1979), the partisans came from the middle classes, the opponents from the old intellectual elite. Phrenology became a pawn in a debate between social classes. The identification of brain functions was seen as a possible basis for a new social policy aimed at giving a chance to anyone blessed with ability, regardless of their social background. To begin with, the squabble was linked to the social interests of the opposing groups; what was at stake was the opportunity offered to individuals to progress in society. It would go on to become an anatomical debate.

The debate on the notion of interest

Relativist analyses formed the subject of a wave of criticism. Some authors rejected the idea that social influence could leave its mark on contents.[11] Others highlighted the fact that the causal relationship between knowledge and social factors had not been demonstrated, but merely claimed. MacKenzie would later be more successful in demonstrating the influence of social interests on the direction chosen by researchers, but this nevertheless fell short of undermining the value of the knowledge produced. He failed to demonstrate how social interests inevitably give rise to certain knowledge contents rather than others. The causal relationship established was weak.

Jurgen Habermas (1971) (in Chapter 8) rejected the relativist analysis because it adopted the instrumentalist attitude that permeated the natural sciences – a manipulative project and knowledge shaped with the general aim of predicting, controlling and, therefore, mastering the universe. He supported a conception of interpretative social sciences (those not seeking social causes) that would improve mutual understanding (communicative and intersubjective rationality).

The ethnomethodologist Steve Woolgar (1981) believed that using the notion of interests was a mistake, as the explanation given by relativists simply replaced nature by society. They took as granted the existence of interests (external causes that determine scientific contents). However, these interests cannot be taken for granted. They have to be negotiated, contested and constructed by the different actors (debate on the factors that prompt individuals to act in one way or another). Therefore, they cannot merely be observed passively as though they already existed.[12] On the contrary, they require just as much of an explanation as scientific content.

Moreover, in their efforts to reveal the true social causes behind apparent scientific rationality, relativists may have produced a sociology of suspicion that denounced obscure motivations and the underlying interests of scientific work.

Ultimately, their theory defined scientists as rational actors seeking to maximise the social interests they represent in science. This conception ignores the fact that scientists are also actors who are submerged in situations that shape and restrict their rationality.

The Empirical Programme of Relativism (Bath)

In 1981, as an extension of the strong programme, Harry Collins (1981) from the University of Bath explained the Empirical Programme of Relativism (EPOR). He put forward a microsociological analysis centred on studies of contemporary cases that were limited to the sociological traits of the social groups studied, without referring to the general social context. He studied local scientific controversies, the manner in which the results were negotiated and the consensuses that explained the production of knowledge. His programme was founded on three principles: (i) symmetrical treatment in the explanation of beliefs (legacy of the strong programme); (ii) identification of the tacit rules of scientific activity

(Kuhn's legacy); (iii) explanation of the mechanisms through which controversies are resolved and their links with the social context. He distinguished three stages:

- Show the **interpretative flexibility** (Box 6.7) of scientific results. Nature always allows several valid interpretations, which are at the root of controversies. The aim is to describe episodes during which scientists attempt to establish the reality of a fact, while experimental results make several interpretations possible.

- Describe both the social mechanisms that limit interpretative flexibility and the construction of consensuses that explain the **controversy closure** (Box 6.8).

- Link the closure mechanisms to social, economic and political structures.

Fluctuating interpretations and the closure of controversies
Collins (1985) focused on the reproduction of experiments ('replication'), a practice that lies at the heart of the scientific process, but also at the root of controversy.

Box 6.7   *Controversial results*

In the 1970s, the physicist Joseph Weber embarked upon the detection of a form of gravitational radiation described in the general theory of relativity. He devised a gravitational-wave detector. In theory, the intensity of the waves was low. This meant that the measurement instrument was crucial and needed to be highly sensitive. However, the results far exceeded expectations. The intensity of the radiation was even greater than the theoretical predictions.

The results announced prompted physicists to check the reproducibility of the results. They built other detectors and launched technical and theoretical discussions on how to produce a sensitive detector specifically for this purpose. However, they did not wish to invest as much effort as Weber had in building such equipment. They decided that more modest detectors would suffice. In the end, the instruments bore little resemblance to each other and it was based on this heterogeneous collection of instruments that the results began to be reproduced. Very soon, physicists were announcing their own results and discrediting those of their colleagues. Every one of them rejected Weber's data, but they did so for different reasons. They produced divergent and varying interpretations.

Replication alone is not enough to close a controversy. Indeed, it may even fuel it further. The protocols, instruments and skill of the experimenter can be called into question and additional parameters brought into play. Controversy increases the number of variables that must be taken into consideration to reproduce the results. When a consensus exists, the result is attributed to a natural cause.

Box 6.8   *Resolving the controversy*

Ultimately, the initial results were rejected while the debate was far from being closed. No empirical data succeeded in making a mark. Replicating the experiment proved problematic. It relied on the working practices of the different individuals, their instruments, procedures, operating methods and tacit knowledge. The results were open to multiple interpretations and the experiments were not enough to resolve the controversy. The experiment itself was at the heart of the discussion. Scientists could not agree on the criteria for a valid experiment. Uncertainty surrounded the validity of the experimental results. The controversy could have continued indefinitely.

However, the controversy did eventually cease when a social process interrupted the experimental circularity. Weber was subjected to a fierce attack by one of his peers, who had uncovered an error in a computer program. The challenger admitted to the sociologist that the error was a minor one, but the force and conviction with which he exploited it allowed him to discredit Weber, despite the fact that he had no comparable instrument at his disposal. The controversy's closure was down to the crucial actions of a single actor, who, in his campaign against Weber, used various means to weaken his opponent's position.

In addition, because Weber's data went beyond the predictions, the entire general theory of relativity, already firmly established in the world of physics, was called into question. Nobody was in a hurry to support Weber. Rejecting his claims and criticising his detector was less risky than confirming his results and questioning the validity of the theory of relativity. The arguments wielded in the debate therefore bore little relation to the process of scientific proof that is supposed to ensure agreement between scientists.

Box 6.9   *Earthworm memory (Collins and Pinch, 1993)*

In this particular case, there were many factors to be taken into consideration and the controversy forced the scientist in question to repeatedly explain his findings. The method he chose to communicate his results proved to be his downfall, however. Indeed, the use of humour to enhance his texts aroused suspicion among his peers, and the apparent simplicity of the experiment was another weakness. Schoolchildren attempted to reproduce the scientist's experiments and then bombarded him with questions. His response was to publish a newsletter: *The Worm Tamer's Gazette*. His popularity with schoolchildren damaged his image in the eyes of his peers, who turned their back on him. A single opponent continued to attack him by continuously raising the intensity of the demands made upon him, up until the moment he retired. Nobody was prepared to pick up the reins and the controversy was never closed; scientists grew tired of the topic and moved on to other problems.

But if the controversy persists, a multitude of causes of all types appear (Box 6.9), and these can condition the results, including, for example, extra-scientific factors such as the image the scientist wishes his/her work to convey.

The existence of a pivotal experiment whose results decide the outcome of a controversy is sometimes put forward as an explanation for its resolution. In the case of the theory of relativity, such decisive proof was apparently put forward. However, the unresolved controversies affecting each piece of evidence indicate that the explanation for the consensus on the theory lies elsewhere (Box 6.10).

Box 6.10   *The theory of relativity and the relativity of decisive proof*

It seems that two decisive elements of proof were produced for the theory of relativity: the experiment performed by Albert Michelson and Edward Morley and the observation of the movement of the stars. Yet, neither of these was historically decisive (Collins and Pinch, 1993). Michelson carried out his experiment in the 1880s, 25 years before Einstein conceived his theory, which took little notice of Michelson's experiments. Another 20 years passed before researchers were able to establish links between Einstein and Michelson, and to present the experiments of the latter as proof of the theory of relativity. However, these results never satisfied Michelson and would lead to a controversy that was never resolved. Indeed, when re-examined in the light of the theory of relativity, they proved to be ambiguous. A more complex version of the experiment was performed in 1925 by a friend of Michelson's, with Einstein's backing. Again, the results were controversial. In 1963, they were still not considered conclusive.

In the meantime, however, the theory of relativity had become so widely accepted in physics that the experiment no longer had any decisive value. Even though the results did not actually confirm the theory, the latter was now beyond questioning. It was considered established. The experiment, which is still used today to support the theory, plays the role of a founding myth rather than that of a decisive result.

As regards the study of star movement, according to Collins and Pinch one could observe a circle of non-independent mutual confirmations between the theoretical predictions and the interpretations of the results observed. None of the many other 'confirmations' of the theory were decisive, but they contributed nonetheless to altering scientific culture.

Belief in the validity of a theory does not depend solely on empirical proof, as the significance of the results are also dependent on whether or not the protagonists are prepared to believe. Again, this relates to the problem of experimental circularity, which is only interrupted by extra-scientific factors.

Tacit agreement
Controversies are less significant when experiments are based on a prior consensus between scientists (Box 6.11).

Core-set or relevant social group
Having demonstrated the fluctuating nature of interpretations and the intervention of social factors (tacit agreement, convention, personal strategy, collective belief) to explain the stabilisation of empirical statements, the third stage of the

Box 6.11   *Consensus prior to the construction of the instrument*

A sensitive experimental device was devised for the detection of neutrinos. This large and costly instrument took a great deal of time to prepare, before being set up in a salt mine. The results were not in line with the theory: there were fewer neutrinos than anticipated. When Weber (in the case of gravitational waves) announced results that contradicted the theory, the response was a defensive one. Nothing of the sort happened this time however, only astonishment and tacit acceptance of the results, although this was not the final word. Nobody attempted to replicate the experiment, nor challenge its validity. Where did the difference lie in this case?

Pinch (1986) showed that the scientist who devised the experiment had previously taken care to consult and work with theoreticians until they approved the experiment's design. The experimental facts were the outcome of a long chain of actions and interactions with colleagues from several disciplines. In addition, when the results were announced, the scientist avoided proclaiming that the theory was false, instead limiting himself to experimental observation, leaving everyone free to make their own interpretation. He left open the question of which theory should perhaps be called into question: that relating to the prediction of neutrino emission, that relating to the behaviour of neutrinos or that governing the detector's design. No single scientific group felt targeted and everyone was free to blame another scientific speciality for the error. The scientist had thus built a consensus around his device. The experimental result formed the subject of tacit consent.

empirical programme suggests linking controversy resolution mechanisms to the social context. It is a question of understanding why interpretation is required. The answer requires analysis of the social groups involved.

During controversies, the actors form 'small networks' or 'core-sets' (Collins, 1985), 'transitional networks' (Edge and Mulkay, 1976) or 'relevant social groups' (Pinch and Bijker, 1987) within which alliances are formed, as well as networks outside the scientific sphere. These are transitional social institutions. Inside them, social influences circulate and are converted. By analysing these groups and their relationships, it is possible to link them to positions in society.

Criticism, Extension and Changes in Direction of the Relativist Programme

The relativist programme marked a turning point in the history of the sociology of the sciences, owing as much to its success as to the criticism and new research programmes it gave rise to.

Criticism

The relativist programme met considerable criticism (Freudenthal, 1984; Ben-David, 1991). Some denied that the approach was new, referring to arguments made previously by Merton and Ben-David. Others denounced the confusion that had arisen between belief and knowledge. The most vehement criticism, however, concerned sociological reductionism (explanation in social terms only, with nature and cognitive factors discarded completely) and its causal model (social forces acting prior to scientific research, rejection of the idea that actors are motivated by a quest for truth). The causal model and, in particular, the principle of covariance (when a variation in the social cause produces a variation in scientific constructs) had apparently not been successfully proven in the cases analysed. Others were unhappy that this form of sociology was reduced to taking into consideration social interests alone, when other sociological interpretations were possible.

Relativism was also the target of invective. The idea that all forms of knowledge were equal was intolerable, and some believed it to be intellectually indefensible and morally dangerous (Freudenthal, 1990). The relativist programme was criticised because it fuelled dissenting movements and criticism of science. It was also criticised by Latour (1987), who wondered what the robustness of hypothetical statements depended on. If they relied on the social groups and conditions under which they were developed, then surely they would vanish as these groups themselves disappeared. However, some hypotheses transcend the context in which they are created. Ben-David believed that the global nature of consensuses should be taken into consideration, that local knowledge is incorporated into a translocal knowledge system (Freudenthal, 1984), the mechanisms of which needed to be described.

Other criticism related to the notions of interest, interpretative flexibility and convention (which lacked a precise definition and were exploited in a way that suggested they had not been fully mastered).

The debate around the symmetry principle deserves special attention. Some sociologists were accused of using the principle asymmetrically and of neglecting to take cognitive factors into consideration (Darmon, 1986). Hess (1997) also highlighted the fact that neutral accounts of controversies, which comply with the symmetry principle, tend to be captured by social groups with the least scientific credibility (the problem of capture). Sociologists' impartiality means that their analyses are invariably adopted by one of the groups involved. Latour also accused the relativist programme of perpetuating asymmetries in the treatment of scientists' output, with a lack of equality between the consideration of natural and social factors. This criticism gave rise to the formulation of a new series of methodological principles, based on the actor-network theory (Callon, 1986), which aims to be both relationist and non-relativist.

Extension of the Relativist Programme to the Study of Technologies

Pinch and Bijker (1987) formulated a programme to analyse the social construction of technologies (SCOT), which followed the different steps of the EPOR programme so as to identify the relevant social groups and the processes through which technologies are stabilised (controversy resolution). The development of a technology is a process of variation and selection:

- **First stage**: show the flexibility in the way the actors interpret the technology and in the way it can be developed. The choices and interpretations made (what is a problem and what is a relevant solution) are evident in technological controversies. They depend on social groups that are easy to define (inventors) or difficult to demarcate (users, who do not necessarily form a homogeneous group, such as 'anti-cyclists' or 'women cyclists', who, for a time, were not supposed to ride bicycles for moral and safety reasons). Once the relevant social groups have been identified, the aim is to understand the role the technical object might play for these groups, the problems each of them might face and the solutions (technical, legal, moral and others) they devise.

- **Second stage**: show how the technological development has stabilised. Stabilisation means that the problem is no longer an issue for the groups concerned. There are two types of stabilisation mechanism: (i) *rhetorical resolution*: the problem has not been resolved, but it has disappeared. The group no longer concerns itself with the problem, perhaps convinced that it no longer exists; and (ii) *practical resolution*: the problem or the solution is redefined. Some solutions (inner tubes in bicycle tyres, for example) to a given problem (vibrations in the bicycle) are rejected (competitive cyclists considered that vibrations were not really a problem). Once the solution has been developed, its supporters nevertheless succeed in having it accepted by showing that it solves another problem (bicycles fitted with inner tubes won races). The definition of the problem corresponding to the solution is transformed.

- **Last stage**: link social groups and stabilisation mechanisms to the rest of society.

Because studies of technical objects sometimes limit themselves to following the innovation process through to the product's technical stabilisation, they leave the impression that things are static and that technical determinism has reclaimed its rights. The technical outcome may incorporate social choices that have a bearing on users. However, interpretative flexibility comes into play in the object's usage, up until its very destruction (where the question of the product's identity and possible uses arises once again). What a machine comprises and is capable of (if it actually works) stems from the interpretations produced alongside the social dynamics at play.

Analysis of Scientific Controversies

Scientific controversies provide a view of the arguments of the various actors involved. Raynaud (2003, p. 8) put forward a restrictive definition of the notion, which excludes squabbles over priorities, debates on the topic of science versus society and occasional disagreements between scientists: 'Persistent public division between members of a scientific community, be they allied or not, who support opposing arguments in the interpretation of a given phenomenon'.

Latour (1987) gave a wider definition: a debate relating, in part, to scientific or technical knowledge that is not yet confirmed or stabilised.

Characterisation of controversies

Raynaud suggested characterising controversies according to eight dimensions:

- **Object**: facts, method principles, theories.

- **Polarity**: number of opposing camps. Controversies generally arise between two rivals, particularly when the controversies grow, but they can involve a greater number of camps depending on the structure of problems and that of the scientific community.

- **Extension**: number of individuals or groups concerned. This may be linked to the degree of commitment shown by the actors involved in the controversy; when extended, the latter requires less commitment on the part of each individual.

- **Intensity**: variable virulence, depending on the controversy and over time, according to the exclusivity of the relationship between rivals and the group's homogeneity. The social structure of scientific groups may explain the dynamics of certain controversies, which can be either strong or imperceptible.

- **Duration**: occasional or prolonged, and apparently limited in cases where the controversy is instituted, that is, with prior definition of stakes, terms and clear indicators of the success or failure of the protagonists involved. When it is not instituted, conversely, it exempts the loser from having to acknowledge his/her failure and allows him/her to reignite the controversy. The level of professionalisation may also have an impact on the dynamics of controversies.

- **Type of forum**: Collins and Pinch (1993) distinguished between the *constitutive forum* (where experimental, theoretical and publication work takes place) and the *unofficial forum* (which includes professional organisations, the recruitment of scientists, popularisation, opinion seeking and so on). Some controversies remain confined to the constitutive forum, while others reach the unofficial forum or move from one forum to the other.

- **Type of recognition**: the controversy can be recognised by just one of the protagonists (unilateral controversy, where the other protagonists consider there to be no controversy) or by several.

- **Type of resolution**: controversial themes can be rejected explicitly (resolved or ended by a formal ruling) or implicitly (exhaustion of some of the protagonists, cost of entering the debate, lack of credibility). Resolution is dependent on various mechanisms: 'negotiation' (Collins, 1985), 'waning interest', 'power struggles', where resources external to the debate are mobilised, adoption of a consensus through adoption of a new perspective, 'development of a convincing argument', and 'negotiation' according to procedures that facilitate the quest for an agreement (Engelhardt et al., 1987). Or, further still: 'redefinition' and 'rhetorical argumentation' (Beder, 1991), 'academic cleansing' (Wallis, 1985) and 'professionalisation' that leads to a change in controversial regime.

In some cases, the actors avoid controversy in order to protect science and its social status (Box 6.12).

Box 6.12   *Non-controversy: the J phenomenon*

At the beginning of the 1920s, Charles G. Barkla studied part of the X-ray spectrum with the aim of analysing an additional series of lines forming the J phenomenon (Wynne, 1976). This eminent physicist, who specialised in X-rays, had already won the Nobel Prize in 1917. His scientific reputation earned him the respect of his peers. However, the J phenomenon, for which he became the spokesman, contradicted quantum physics. Would his results be rejected, as others had been in the past, because they called into question the foundations of his colleagues' work, or would they be met with enthusiasm?

In actual fact, other physicists failed to react: there was no official rejection, no polemic and no enthusiasm. Nobody believed in the J phenomenon, but nor did anyone trouble Barkla, who pursued his work and went on to oversee a number of PhD theses, all relating more or less to the study of the J phenomenon. Moreover, his PhD students had no trouble either defending their theses or obtaining jobs in academic research. However, once they were no longer supervised by Barkla, they ceased to profess their belief in the J phenomenon. His social standing allowed him to pursue his work, but his divergence from the dominant belief meant that his work remained unrecognised. He was never accused of fraud or irrationality, because the image and social status of science, as embodied by this Nobel Prize winner, were at stake.

Controversy as an instrument of analysis

Controversy exists because nature and empirical evidence can be interpreted in several ways. From a relativist perspective, its analysis uncovers the flexibility of interpretations and closure processes. From a Latourian perspective, it reveals the processes whereby knowledge is stabilised through an accumulation of asymmetries that must be identified. In both cases, controversies provide sociologists with valuable observation points. They reveal the different actors, the way in which they construct facts and theories, the nature of the arguments used and their contingency. Monitoring a controversy makes it possible to cross-examine the elements that contribute to the result for as long as they are visible. Indeed, before becoming black boxes, these elements are assessed, negotiated, transformed, tested and consolidated. The ideal situation for an observer is to be present during the controversy, as once this has been resolved, a great deal of information is lost, in particular those negotiations that were never recorded. The interpretations made by the different actors a posteriori are barely usable reconstructions and rationalisations (Box 6.13).

Up until their resolution, technological controversies make it possible to define contents, as well as the limits between what has been established and what has not, between what is feasible and what is not, between what is research and what is application, but also the dividing lines between technical content and social

---

Box 6.13   *The novelty controversy*

When Mendel published his hybridization results, they were not a new discovery as such, as they followed in the tradition of hybridization. This could hardly be described as a scientific revolution. However, 40 years later, when Carl Correns and Hugo De Vries quarrelled about being the first to demonstrate the theory of heredity, Correns buried his rival by proclaiming that Mendel was the precursor and originator of this fantastic scientific revolution. 'Being a discovery' is not an intrinsic property, but is relative to the knowledge and problems of the period in history (Brannigan, 1981).

Box 6.14   *Method: analysis of a controversy*

1 **The actors and the means at their disposal for the production of knowledge**: taking as a starting point the arguments of scientific actors, the practical resources allowing them to produce knowledge must be identified, along with the scientists, technicians, institutions and instruments behind the data, theories, instruments and arguments. It is also important to identify their colleagues, including those from other disciplines, and the means at their disposal (instrumentation, theories and data). The controversy takes place either between actors within scientific communities or between individuals from different communities. Analysing the controversy involves mapping the relationships between these actors.

2 **Actors from the secondary network**: scientists liaise with sponsors, investors, foundations, industrial firms and those who use their results, from whom they receive resources and to whom they supply a case statement (a research project, a professional or popularised publication, a research report, a demo and so on). These actors (allies, spokespeople, opponents, press and the public) are often numerous and act according to interests, values, forms of organisation and world views that differ from those of researchers. The controversy can extend beyond the realm of science into this secondary network or vice versa. Again, the various relationships must be mapped.

3 **Shaping the controversy**: the controversy is often shaped by the actors themselves, who attempt to resolve it by means of an academic review, a colloquium, a measurement campaign, legal proceedings, a televised debate or a parliamentary debate. The way a controversy is shaped governs the dynamics at play. It is important to report on this because it shows how the protagonists themselves analyse the controversy.

4 **The dynamics of the controversy**: reporting on the way in which the controversy evolves (changes in contents and arguments, amplification or fading, entry or departure of new actors, specialisation or popularisation, polarisation or consensus, intensification or dilution and so on).

---

contexts (Callon, 1980). Their analysis (Box 6.14) leads to reports presenting the points of view of the different actors and the relationships they build, as well as the uncertainty surrounding the outcome. This makes it possible to understand the impact of contingencies on the production and assessment of knowledge.

An example: the Pasteur–Pouchet controversy

In his analysis of the controversy that pitched Pasteur against Felix Archimède Pouchet, on the topic of spontaneous generation, Latour (1995) suggested that the differences that explain what is true and what is false should not be presupposed. Instead, it is important to report on the dynamics of the controversy, the arguments put forward and the resources mobilised by the protagonists. The academy ultimately sided with Pasteur, despite the fact that his empirical proof was inconclusive. This can be explained by the intervention of other factors: prejudice and beliefs, the social standing of the protagonists, the ideological and political context, as well as rhetoric and manoeuvring. Latour drew up a list of asymmetries and was thus able to draft a report on Pasteur's victory.

Raynaud (2003) contested his conclusions by showing that the asymmetries were not as clear as they had initially appeared, and that other asymmetries had not been taken into account and the bias could be reversed. He suggested that greater exhaustiveness was required in the analysis of asymmetries, as the bias depends on which elements are taken into consideration.

Relationist Extension: Callon–Latour

As they reviewed, criticised and radicalised the relativist programme within the sociology of the sciences, Callon (1986) and Latour (1995) defined a new series of basic principles for the study of science and technology:

- **Bloor's symmetry principle (ensuring that the explanation is symmetrical, regardless of the outcome)**: avoid automatically introducing a greater degree of reality or rationality in scientific statements compared with other statements. It is not acceptable to base one side of the argument on nature, logic or scientific methods and the other on social and psychological factors. Similarly, social factors should be used to explain both knowledge that is recognised as valid and beliefs that are judged to be false.

- **The principle of symmetry between nature and society**: report, in the same terms, on technical and social aspects and on local events and circumstances. However, the terms of nature and society explain nothing in themselves and they too must be explained:

  > Because the resolution of a controversy is the cause of a stable representation of nature, rather than being its consequence, one can never use the consequence or the state of nature to explain how and why a controversy was resolved. . . . Because the resolution of a controversy is the cause of a stable society, one cannot use the state of society to explain how and why a controversy was resolved. (Latour, 1987, p. 426)

- **The principle of agnosticism in treating actors' discourse, regardless of the subject in question, be it natural or social**: do not favour any of the opinions expressed by the actors studied. Relativist sociology avoided passing judgement on the way in which scientists analyse nature. This principle must be extended to their discourse on society. The relativists acknowledge the right of scientists to trigger controversies on questions relating to nature (relativism with respect to nature), but they do not accept that these controversies may be extended to society (on which they are the experts: no relativism with respect to society). They grant society a decisive role that they refuse to give to nature or logic. Yet, disagreements between actors also relate to the definition of society and its actors. Failure to respect this principle of agnosticism leads to a number of problems: (i) *stylistic*: reports ignore the discussions held by actors on the topic of social structures, which are erased from a part of their constructions; (ii) *theoretical*: controversies between sociologists on the explanations to be used are as interminable as those of the scientists they study. Because every element of knowledge on nature and society is as debatable as any other, these elements cannot be made to play different roles in the analysis; and (iii) *methodological*: observers who are unaware that the identity and characteristics (interests, intentions, forces and so on) of the actors are permanent subjects of discussion, are liable to take these actors for granted, when the reality is in fact more problematic. It is therefore important to record any uncertainties relating to this identity in the event that it is controversial.

- **The principle of symmetry between the human and the non-human**: sociologists reject the idea that non-human entities (including instruments) should have a say because they do not allow nature to be granted a favourable position in the explanation. However, these entities cannot be carved into shape and forced to serve our purposes at will. Their presence, movement, expression, action and reaction must be observed in the same way as those of humans.

- **The principle of circumstances and associations**: nothing escapes contingency, negotiation, interaction, situations and circumstances, be it the interpretation of results, the reproduction of experiments, or the production of facts or criteria allowing the relevance of a piece of evidence to be judged. The aim is to apprehend the circumstances and events that take place, to understand the different interactions without imposing a pre-established analysis grid or systematic distinctions, and to follow the movements of entities. For this principle to be applied, the manner in which the actors define and associate the different elements must be taken into account. This involves drawing up an inventory of the categories used, the entities mobilised and the relationships into which they enter, in addition to their calling into question.

Callon (1986) put these principles into practice in his analysis of the activities of a group of marine biologists who were endeavouring to produce new knowledge about scallops. This was one of the founding texts of the 'sociology of translation' and the 'actor-network theory' (ANT).

Contextual Knowledge

To complete this analytical overview of the sociology of scientific content, let us finish by examining recent work, which highlights the contextual nature of knowledge.

The context of knowledge is, first and foremost, that of an 'epistemic culture', which varies depending on the discipline (Knorr-Cetina, 1999). High-energy physics, for example, is linked to major organisations and research practices that exist outside human time and space while molecular biology, on the other hand, is more firmly attached to human space and time. The former works with the signs and traces produced by instruments, the latter with objects manipulated experimentally. High-energy physics operates as a closed social and epistemic community, which works on a range of objects defined by complicated technology. Its attention is focused not on observing the world, but on the instrument and its personality (age, illnesses, life expectancy, traits of character, reflection of the person who designed it), which governs the status of scientists and the relationships between them. Social segregation is common within the discipline and is dependent on the efforts made to launch research programmes. Physics is multicultural and highly structured: instrument manufacturers who are familiar with the manipulation of gases, liquids and circuits; theoreticians preoccupied with consistency and quantification; experimenters obsessed with measurement. Between them, there are wide 'border trading zones' (Galison, 1997). Trading with society, however, is restricted to the bare minimum required to build and operate instruments and experiments. In molecular biology, however, the research system is geared towards the objects of nature, which are transformed into molecular machines. Researchers, together with their corpus and tacit knowledge, occupy a central position where individualism flourishes. Projects are personal and dependent on the gathering of local human and material resources, spread between multiple knowledge 'production sites'. Relationships between scientists and society also differ from one discipline to the next. Society, in particular the socioeconomic world and the public, is unafraid to make its opinion heard on the subject of molecular biology, while any comment on high-energy physics is usually barely a murmur.

The contextualisation of knowledge depends on the degree to which society's preoccupations are taken on board in each area of research (Nowotny et al., 2001). High-energy physics research depends on cooperation with political and industrial actors who have an influence in the decision-making process that accompanies the construction of large particle accelerators. Links with the scientific contents produced remain weak, however, despite the weight of political and military interests (Krige and Pestre, 1997); the almost tribal relationships that develop between researchers has had more of an influence on the direction taken by contents than their relationships with sponsors. The astronomical cost of instruments (and the reticence this generated on the political front at a time when the Cold War was beginning to wane) is the only factor to have led scientists to explore less costly concepts and to cooperate with other disciplines. The intensity of collaboration around very large instruments and long-term research programmes, the strength of internal social ties and the precedence given to understanding the fundamental elements of the universe and the scientific challenges thrown up by previous discoveries, all serve to explain that the knowledge produced has little do with context. It is, however, affected by cost constraints, the design capabilities of manufacturers and the political equilibrium of international cooperation. The design of particle accelerators and their geographical location have been impacted by these factors, although the influence of the latter on knowledge content remains very indirect. The same is true in cases where national research programmes have been set up. Run by scientists who interpret or even ignore society's messages, the influence of context on these programmes remains fairly limited.

In other situations, the contextualisation of knowledge is much stronger. The explanation lies less in the fact that scientists pursue objectives defined by the outside world, than in the intensity of their exchanges with society and the changes in perception that these prompt. These exchanges have an impact on the research subjects and problems selected, on the constraints considered during the problem-posing process and from a methodological perspective. Contextualisation also relies on the fact that scientists and other social actors develop shared viewpoints of the world, its problems and possible solutions, creating a sort of interculturality (Yearley, 1996). Contextualisation is more apparent in the case of technological projects, making it more difficult to differentiate between internal and external factors. Technical design is governed by multiple standards defined by national and international bodies, which represent the requirements and priorities of society, states and socioeconomic actors. The involvement of civil society in scientific and technical issues (public debate, role of patients' associations) (Rabeharisoa and Callon, 2002) also contributes to the contextualisation of knowledge through the joint definition of orientations, finalities, themes, methods and the relative value of knowledge. The map of the human genome falls into this category of partially contextualised knowledge.

According to Nowotny et al. (2001), there is no longer an irreducible epistemological core, comprising of cognitive values and a scientific ethos. The sciences are now populated by multiple forces, qualified in the past as extra-scientific and dispersed in multiple knowledge production contexts. The decline of normative and cognitive authority once associated with science can be explained by divergences between the interests of those who produce knowledge in a way that is geared towards its normalisation and consistency, in spite of the diversity of usage contexts, and the interests of users who prefer knowledge that builds on what they have already learned.

Conclusion: Models for the Study of Science

The intellectual panorama of the analysis of science comprises a series of analytical models, some of which were presented earlier:

- The **naturalist and positivist model of science**: scientific statements are 'dictated' by nature, and it is important to listen to nature while discarding any prejudices and taking heed of the illusions that threaten scientists. Resorting to the *right* scientific method, to logic and to the appropriate instrumentation makes it possible to 'uncover', 'reveal' and 'transcribe' the laws of nature. This conception of science is sometimes applied by sociologists themselves *vis-à-vis* society when it comes to describing the laws operating within the latter.

- The **internalist model**: scientific statements stem from earlier ideas. The ideas explaining the development of the sciences follow a logical progression.

- The **conceptual contextualisation model**: scientific statements are born from previous ideas, including those relating to philosophy, art and society. However, the production of scientific ideas is confined to the world of science, even if it is stimulated by external sources. Ultimately though, institutional or societal causality is not acknowledged in the explanation (see Koyré, 1958 on the history of science or Bachelard, 2002 on the philosophy of science).

- The **sociological conditioning model**: the creation and fructification of statements rely on favourable social conditions. In philosophy, Lakatos (1978) defended the idea that the best research programmes inevitably rise to the fore, but are delayed to varying extents by social conditions. In sociology, Ben-David (1971) (in Chapter 1) suggested that ideas are carried forward by those groups that choose to develop them and nothing guarantees that even the best ones will be successful. The expansion of science depends on social conditions.

- The **social relativism model**: scientific statements are social products determined by the beliefs of social groups and by social structures (the sociological reductionism model presented in the previous chapter).

But the panorama of sociological analyses does not stop there. The following chapters will provide an overview of other models:

- The **(social) constructivism model**: scientific statements are social products that can be explained by the processes resulting from these constructs.

- The **'transversality model of scientific activity'** and the socio-epistemological model: these combine: (i) cognitive conditioning, linked to the intellectual paths of individuals, forms of reasoning (for example, quantification) and working practices (metrology, standardisation and so on). (ii) socio-strategic conditioning, linked to professional reputation, social position and power strategies. These elements form a framework that guides the researcher. This model draws from the constructivist approach, but also strives to do justice to a conception of science that cannot be boiled down to an economic or political struggle to mobilise resources and extend its influence. It also covers numerous preconceived factors and dimensions (cognitive and social motivations and conditioning) (Feltz, 1991, Gingras, 2000 (in Chapter 7), Kreimer, 1997 (in Conclusion) Shinn and Ragouet, 2005.

- The **ethnomethodological model**: scientific statements are local products that emerge from the interactional dynamics between the protagonists of a situation who share a number of skills, in particular linguistic ones. They are practical, contingent and situated accomplishments. This model does not claim to cover the scientific dynamics that transcend the local situations studied (Lynch, 1985).

- The **actor-network model** (Callon, 1986, Latour, 1987): scientific statements are sociotechnical products. They rely on different sociotechnical networks, some more extensive and robust than others (see Chapter 7).

The fact that content is now considered in the study of scientific dynamics has also led to a shift in the level at which analyses are performed. They have become more microsociological and their conclusions are difficult to apply across the board. It has therefore become necessary to increase the number of areas of investigation and address each field, discipline, institution, nation and era individually. Gradually, a corpus of case studies has been assembled, but their comparison remains problematic, so variable are the lines of questioning and methodologies used by the authors. There is still a sizeable challenge in terms of developing a sociology of the sciences that takes into account contents and equips itself with the resources needed to construct mid-range models and theories.

Box 6.15   *Going further: the sociology of the sciences at the end of the twentieth century*

The end of the twentieth century witnessed the rapid development of social studies on science and technology, which was not entirely unrelated to growing questioning about the role of science in society. As of the mid-1960s, research centres were being set up in Europe around the question of the relationship between 'science and society', such as the Science Policy Research Unit (SPRU) at the University of Sussex, in 1965, and the Science Studies Unit at the University of Edinburgh (David Edge), in 1966. Others followed with the aim of studying the relationship between science, technology and society (STS). An example is the Centre for the Sociology of Innovation at the École des Mines de Paris, which focused on large-scale industry and would later welcome Callon and Latour.

In 1969, MacLeod and Edge launched a publication which, in 1974, was christened *Social Studies of Sciences* and was one of the main periodicals in the field. In 1978, the periodical *Scientometrics* was launched to present more quantitative studies. Collaborations were formed on the initiative of the British research centres via the PAREX (Paris–Sussex) association, which led to the creation, in 1981, of the European Association for the Study of Science and Technology (EASST). Over the course of the 1960s, more researchers entered the field, notably Barnes, Bloor, Mackenzie and Pickering.

Over in the United States, in 1957, Merton called for the development of the sociology of science within the American Sociological Society, the American Sociological Association (ASA) and the International Sociological Association (ISA). He had a dominant influence in the field, but young scientists were calling for more 'radical change'. They founded a new society for the study of the social aspects of the sciences: the Society for Social Studies of Science (4S), in which Merton's influence gradually faded, to be overtaken by the constructivist, feminist and culturalist approaches. Indeed, in 1996, Stephen Cole criticised the socio-constructivist approach's monopolisation of the 4S. Merton considered that a large proportion of 'constructivist' works made little sense, but he refrained from pursuing his argument.

In addition, in both the United States (Cornell, Harvard, MIT and so on) and Europe (Lund, Gothenburg, Bielefeld and Conservatoire National des Arts et Métiers (CNAM) in Paris), various cross-disciplinary programmes emerged relating to STS, Science Policy and Social Studies of Sciences. A number of periodicals and collections were created, including *Pandore*, which was edited by Callon, Latour and Phillipe Mallein.

This gradually emerging new research community was heterogeneous in terms of the fields it covered and the approaches it took. Dubois (2001) separated it into four groups:

- **The Mertonian group**, promoting the sociology of scientific institutions.

- **The cross-disciplinary PAREX group**, promoting empirical research reflecting the plurality of scientific dimensions and space.

- **The 'strong programme group'**, including individuals from the Universities of Bath and Edinburgh (Barnes, Bloor, Collins, MacKenzie, Pickering, Pinch) and arguing for empirical research on social interests and the local cultural systems that govern scientists.

- **The (socio-)constructivist group** including several research movements that shared their intention to focus on concrete practices, analysed in situ (see Chapter 7). Institutional, political, economic and cognitive contexts are only taken into account through their local mediations. The works of Knorr-Cetina, Lynch and the earlier works of Latour and Woolgar can be included in this group.

Dubois analysed how relations between these two groups have progressed over time: the influence of the Mertonian group has gradually weakened to the benefit of the 'strong programme' group and the constructivists, from which the 'actor-network' approach has gradually broken away.

Box 6.16   *Relativism / rationalism: the great debate*

The sociology of science is engaged in a debate between **relativism and rationalism**. This is an age-old conflict. From Plato to the positivists, rationalists have always considered there to be a common pool of immutable reality that is accessible to reason. Conversely, relativists consider that things change and that the truth is neither unique nor universal, that it varies according to the observer and their society.

With regard to science, the debate revolves around the **notions of proof and consensus**. Proof is grounded in logic and reason, whereas the basis of consensus is social.

In the opinion of **rationalists** such as Larry Laudan, Lakatos and Hollis, proof that is correct is accepted automatically, or at least it is by those who are competent and without prejudice. It leads to a consensus and derives its power from the structure of reasoning and from its links to nature. If a consensus is not achieved, it is the result of a lack of sufficient information, blinkered ideological prejudices and resistance to change. Consensus can be explained by the empirical and logical value of proof, non-consensus by external, psychological and sociological factors.

Relativists, including Barnes, Bloor and Collins, refused to presume the existence of absolute and universal rationality criteria. In other words, what is accepted as a valid or rational argument varies depending on the context. What is recognised as proof by one group is not necessarily accepted by another. *Proof is relative.* It depends on the local context and on a belief system. The observer cannot, therefore, express an opinion on what is or is not rational.

A **consensus** exists from the moment the members of a group recognise an argument as complying with its own proof criteria. The criteria themselves depend on the group and can be explained sociologically. *Consensus is a social phenomenon.* It is the result of beliefs whose obscure social origins may lead one to believe that they are objective truths. Science and the notions of proof, reason, validity and objectivity are beliefs and categories specific to a social group. Scientific theories are reliant on the social consensuses they achieve.

This debate between rationalism and relativism was the basis for a significant proportion of the discussions that took place at the end of the twentieth century. It would, nevertheless, be useful to examine the thinking of the different authors more closely, as very few of them put forward analyses that are as stereotypical as the above summary of the rationalist and relativist positions.

Notes

1 In the Popperian sense, the justification of hypothetical statements is not social justification of scientific activity as such, it is the scientific validation of hypotheses (a posteriori identification of intrinsic rationality).

2 The Paris–Sussex Association: Gérard Lemaine, Roy MacLeod, Michael Mulkay, Terry Shinn, Peter Weingart and Richard Whitley.

3 See Barber and Fox (1958), Thill (1973), Latour and Woolgar (1979 [1986]), Feltz (1991), Gooding (1992), Vinck (1992).

4 Hypotheses tend to be linked to one another and so their experimental invalidation calls into question whole sets of theories (Duhem–Quine thesis).

5 The notion of convention, inspired by the work of Duhem, provides fuel for the 'sociological framework theory', one of the legacies of Kuhn's work.

6 Latour (1996) (in Chapter 7) preferred to defend the idea of 'interobjectivity': fact is the product of interdefinition between the natural elements perceived via instruments, the signs produced by these instruments, the concepts developed by scientists to produce and read phenomena, and the negotiations and conventions binding them.

7 Kuhn, however, did not see his popularity among sociologists in a positive light. He frowned upon the adoption of his ideas and the relativist conclusions they gave rise to, strongly denying that he had ever had such intentions.

8 According to the analysis of Margaret Masterman (1970), who was a disciple of Kuhn.

9 The notion of incommensurability has been criticised. Saying that two theories are incommensurable is tantamount to saying that it is impossible to translate one into the other. However, it was Popper's contention that even languages as different as Chinese

and English can be translated into each other, and individuals who speak one of the languages can learn to master the other. Referring to Quine, Kuhn responded that several translations are possible, as no translation can ever be perfect. Concepts are always linked to their context.

10 When it came to choosing between the charm and colour models of high-energy physics (Pickering, 1981), the first model came out on top because it could more easily be incorporated into the current practices of the different groups of physicians.

11 Relativist sociology might reply: if one cannot see the mark made by society, it is because this mark has been erased. The credibility of a scientific claim relies on the absence of evidence of any social motivation that might otherwise be used against the scientist to discredit his/her scientific pretensions.

12 Barnes (1981) replied that one must not confuse the interests at work, as reconstructed by the sociologist during the analysis, with the interests that scientists themselves perceive and imagine.

Recommended Reading

References appearing in other chapters: Bijker et al. (1987) in Chapter 7.

Barnes, B. (1974), *Scientific Knowledge and Sociological Theory*, London: Routledge & Kegan Paul.

Bloor, D. (1976), *Knowledge and Social Imagery*, London: Routledge.

Callon, M. (1986), 'Some elements of a sociology of translation: domestication of the scallops and the fishermen of St Brieuc Bay', in J. Law (ed.), *Power, Action and Belief: A New Sociology of Knowledge*, London: Routledge & Kegan Paul, pp. 196–233.

Collins, H. (1985), *Changing Order: Replication and Induction in Scientific Practice*, London: Sage.

Collins, H. and Pinch, T. (1993), *The Golem: What Everyone Should Know about Science*, Cambridge: Cambridge University Press.

Edge, D. and Mulkay, M. (1976), *Astronomy Transformed: The Emergence of Radio Astronomy in Britain*, New York and London: John Wiley & Sons.

Engelhardt, H.T. and Caplan, A. (eds) (1987), *Scientific Controversies: Case Studies in the Resolution and Closure of Disputes in Science and Technology*, Cambridge: Cambridge University Press.

Hess, D. (1997), *Science Studies: An Advanced Introduction*, New York: New York University Press.

Knorr-Cetina, K. (1999), *Epistemic Cultures: The Cultures of Knowledge Societies*, Cambridge, MA: Harvard University Press.

Kuhn, T. (1962), *The Structure of Scientific Revolutions*, Chicago, IL: University of Chicago Press.

MacKenzie, D. (1981), *Statistics in Britain 1865–1930: The Social Construction of Scientific Knowledge*, Edinburgh: Edinburgh University Press.

Pinch, T. (1986), *Confronting Nature: The Sociology of Neutrino Detection*, Dordrecht: Reidel.

References

References appearing in other chapters: Bernal (1939), Grossetti and Mounier-Kuhn (1995), Polanyi (1958) in Chapter 1; Lemaine et al. (1969), Merton (1938) in Chapter 2; Cole and Zuckerman (1975), Krige and Pestre (1997), Mullins (1972) in Chapter 3; Nowotny et al. (2001) in Chapter 4; Bourdieu (1975) in Chapter 5; Collins (1974), Latour and Woolgar (1979 [1986]), Lynch (1985), Thill (1973), Vinck (1992) in Chapter 7; Latour (1987) in Chapter 8.

Bachelard, G. (2002), *The Formation of the Scientific Mind: A Contribution to a Psychoanalysis of Objective Knowledge*, Manchester: Clinamen. [Original edition in French, 1938.]

Barber, B. and Fox, R. (1958), 'The case of the floppy-eared rabbits: an instance of serendipity gained and serendipity lost', *American Journal of Sociology*, **64** (2), 128–36.

Barnes, B. (1981), 'On the "Hows" and "Whys" of cultural change', *Social Studies of Science*, 11 (3), 481–98.

Barnes, B. (1982), *T.S. Kuhn and Social Science*, New York: Columbia University Press.

Beder, S. (1991), 'Controversy and closure: Sydney's beaches in crisis', *Social Studies of Science*, **21**, 223–56.

Ben-David, J. (1991), *Scientific Growth: Essays on the Social Organization and Ethos of Science*, Berkeley, CA: University of California Press.

Brannigan, A. (1981), *The Social Basis of Scientific Discoveries*, Cambridge: Cambridge University Press.

Callon, M. (1980), 'Struggles and negotiations to decide what is problematic and what is not: the socio-logics of translation', in K. Knorr-Cetina, R. Krohn and R. Whitley (eds), *The Social Process of Scientific Investigation*, Dordrecht: Reidel, pp. 197–220.

Collins, H. (1981), 'Stages on the empirical programme of relativism', *Social Studies of Science*, 11(1), pp. 3–10.

Darmon, G. (1986), 'The asymmetry of symmetry', *Social Science Information*, **3** (25), 743–35.

Dean, J. (1979), 'Controversy over classification: a case study from the history of botany', in B. Barnes and S. Shapin (eds), *Natural Order: Historical Studies of Scientific Culture*, London and, Beverly Hills, CA: Sage, pp. 211–30.

Dubois, M. (2001), *La nouvelle sociologie des sciences*, Paris: Presses Universitaires de France.

Duhem, P. (1906), *La théorie physique: son object et sa structure*, Paris: Chevalier R. Riviè [repr. Paris, Vrin, 1981].

Elkana, Y. (1968), *The Discovery of the Conservation of Energy*, London: Hutchinson.

Feltz, B. (1991), *Croisées biologiques. Systémique et analytique. Écologie et biologie moléculaire en dialogue*, Brussels: Ed. CIACO.

Feyerabend, P. (1975), *Against Method: Outline of an Anarchistic Theory of Knowledge*, London: Humanities Press.

Fleck, L. (1935 [1979]), *The Genesis and Development of a Scientific Fact*, Chicago, IL: University of Chicago Press.

Forman, P. (1980), 'Weimar culture, causality, and quantum theory, 1918–1927. Adaptation by German physicists and mathematicians to a hostile intellectual environment', in C. Chant and J. Fauvel (eds), *Darwin to Einstein: Historical Studies on Science and Belief*, London: Open University Press; Longman.

Freudenthal, G. (1984), 'The role of shared knowledge in science: the failure of the constructivist programme in the sociology of science', *Social Studies of Science*, **14**, 285–95.

Freudenthal, G. (1990), 'Science studies in France: a sociological view', *Social Studies of Science*, **20**, 353–69.

Galison, P. (1997), *Image and Logic. A Material Culture of Microphysics*, Chicago, IL: University of Chicago Press.

Gieryn, T. (1978), 'Problem retention and problem change in science', in J. Gaston (ed.), *Sociological Inquiry: The Sociology of Science*, San Francisco: Jossey-Bass, pp. 96–115.

Gieryn, T. (1995), 'Boundaries of sciences', in S. Jasanoff, G. Markle, J. Peterson and T. Pinch (eds), *Handbook of Science and Technology Studies*, London: Sage, pp. 393–443.

Gooding, D. (1992), 'Putting agency back into experiment', in A. Pickering (ed.), *Science as Practice and Culture*, Chicago, IL: Chicago University Press, pp. 65–112.

Hessen, B. (1931), 'The Social and economic roots of Newton's Principia', in N. Butharin, *Science at the Crossroads*, London, pp. 151–212.

Hollis, M. (1988), *The Cunning of Reason*, Cambridge: Cambridge University Press.

Hollis, M. and Lukes, S. (eds) (1982), *Rationality and Relativism*, Cambridge, MA: MIT Press.

Horton, R. (1973), *Modes of Thought*, London: Faber & Faber.

Koyré, A. (1958), *From the Closed World to the Infinite Universe*, New York: Harper.

Lakatos, I. (1978), *The Methodology of Scientific Research Programmes: Philosophical Papers Volume 1*, Cambridge: Cambridge University Press.

Latour, B. (1995), 'Pasteur and Pouchet: the heterogenesis of the history of science', in M. Serres (ed.), *History of Scientific Thought*, London: Blackwell, pp. 526–55.

Law, J. (1976), 'The development of specialties in science: the case of X-ray protein crystallography', in G. Lemaine, R. McLeod, M. Mulkay and P. Weingart (eds), *Perspectives on the Emergence of Scientific Disciplines*, The Hague: Mouton, pp. 123–51.

Lemaine, G., Lécuyer, B., Gomis, A. and Barthélemy, C. (1972), *Les Voies du succès. Sur quelques facteurs de la réussite des laboratoires de recherche fondamentale en France*, Paris: GERS.

Lukes, S. (1973), 'On the social determination of truth', in R. Horton (ed.), *Modes of Thought*, London: Faber & Faber, pp. 230–48.

Mannheim, K. (1967), *Essays on the Sociology of Culture*, London: Routledge & Kegan Paul.

Masterman, M. (1970), 'The nature of a paradigm', in I. Lakatos and A. Musgrave (eds), *Criticism and the Growth of Knowledge*, Cambridge University Press, pp. 59–89.

Mendelsohn, E. and Elkana, Y. (1981), *Sciences and Cultures. Anthropological and Historical Studies of the Sciences*, Sociology of the Sciences Yearbook, Dordrecht: Kluwer.

Mulkay, M. (1980), *The Sociology of Science in East and West*, London: Sage.

Pickering, A. (1981), 'The role of interests in high-energy physics: the choice between charm and colour', in K.D. Knorr-Cetina, R. Krohn and R.D. Whitley (eds), *The Social Process of Scientific Investigation*, Sociology of the Sciences, Vol. 4, Dordrecht: Reidel, pp. 107–38.

Pinch, T. and Bijker, W. (1987), 'The social construction of facts and artifacts, or how the sociology of science and the sociology of technology might benefit each other', in W. Bijker, T. Hughes and T. Pinch (eds), *The Social Construction of Technological Systems*, Cambridge, MA: MIT Press, pp. 17–50.

Popper, K. (1972), *The Logic of Scientific Discovery*, London: Hutchinson (1st edn 1959).

Quine, W. (1974), *Two Dogmas of Empiricism, From a Logical Point of View*, Cambridge: Cambridge University Press.

Rabeharisoa, V. and Callon, M. (2002), 'The involvements of patients' associations in research', *International Social Science Journal*, **171**, 57–65.

Raynaud, D. (2003), *Sociologie des controverses scientifiques*, Paris: Presses Universitaires de France.

Scheler, M. (1926), 'Die Wissenformen und die Gesellschaft' ['The forms of knowledge and society'], *Reedt in Collected Works*, vol. 8, Berne: Francke Verlag, 1960.

Shapin, S. (1979), 'The politics of observation: cerebral anatomy and social interests in the Edinburgh phrenology disputes', in R. Wallis (ed.), *On the Margins of Science: The Social Construction of Rejected Knowledge*, Sociological Review Monographs, Vol. xxvii, Keele: Keele University Press, pp. 139–78.

Shinn, T. and Ragouet, P. (2005), *Controverses sur la science. Pour une sociologie transversaliste de l'activité scientifique*, Paris: Raisons d'agir.

Sorokin, P. (1957), 'Social and cultural dynamics: a study of change in major systems of art, truth, ethics, law and social relationships', Boston, MA: Extending Horizons Books, Porter Sargent.

Wallis, R. (1985), 'Science and pseudo-science', *Social Science Information*, **24** (3), 585–601.

Wittgenstein, L. (1922 [1961]), *Tractatus Logico-Philosophicus*, New York: Humanities Press.

Wittgenstein, L. (1953 [2001]), *Philosophical Investigations*, London: Blackwell.

Woolgar, S. (1981), 'Interests and explanation in the social study of science', *Social Studies of Science*, **11** (3), 365–94.

Wynne, B. (1976), 'C.G. Barkla and the J phenomenon: a case study in the treatment of deviance in physics', *Social Studies of Science*, **6**, 307–47.

Yearley, S. (1996), *Sociology, Environment, Globalisation*, London: Sage.

Znaniecki, F. (1940), *The Social Role of the Man of Knowledge*, New York: Columbia University Press.

7 Scientific practices

Until now, two approaches have structured the sociology of sciences. The first concerns sciences within institutions, organisations or systems of exchange. The second reports on what scientists produce by analysing the influence of social processes on the content of scientific knowledge. However, neither of these two approaches looks close up at what scientists do in their work on a day-to-day basis. By looking at controversies, sociologists are held back at a level of discourse and conceptual production. Practices have remained secret or been reduced to their sociological causes. For Shapin (1979) (in Chapter 6), material techniques are simply a materialisation of the interests of groups that are in competition with each other. Collins (1974) insists on experimental practice, but he does so within the framework of a theory of experimental circularity, to show that experimental regression can only be stopped by social factors.

There are few authors who take an interest in concrete practices. Wittgenstein, with his notion of the game of language, established the basis for their study. Kuhn emphasised the importance of 'reasonable agreements' in laboratory practice. Fleck and Polanyi drew attention to practices, instruments, experimental mechanisms and technicians as well as to tacit know-how. Ravetz believes that we cannot achieve academic excellence with formal principles, but only through day-to-day practice. He writes: 'Although tools are only auxiliaries of the advancement of scientific knowledge, their influence on the directions of work is important and often decisive' (Ravetz, 1972, p. 89).

Looking at concrete, ordinary, in situ practices is exactly what researchers have been doing since the 1970s (Box 7.1). They talk about going to see what is happening on the field, in laboratories, of watching scientists at work (Latour (1979) wrote an article entitled 'Go and see') and report on the process of creation of knowledge.

Box 7.1 *Authors of laboratory studies*

The Belgian physicist and philosopher George Thill was something of a precursor with his analysis of high-energy physics, *La Fête scientifique* (1973). Involved in the analysis of the results of a particle collision experiment, he describes the work in its epistemic, organisational and anthropological dimensions. He sees this work as a parenthesis in life (a 'scientific feast') where the normal rules of life are turned upside down. Scientific practice is an action which invents an intrinsic utopia with a rational course. Published in French, his work was to have little influence on the sociology of sciences. Feltz (1991), his young colleague, published a comparative analysis of two laboratories (cellular biology and aquatic ecology) where he combines epistemological and sociological analysis.

In 1977, Gérard Lemaine's team in Paris (working in conjunction with British and German sociologists since the beginning of the 1970s), reconstituted the evolution of a laboratory in the neurophysiology of sleep (Lemaine et al., 1977). Their study reports on the 'ecology of choices' of the laboratory, taking into account training of its members, their epistemological starting points and the technical difficulties that they encountered. The team analysed the strategies of researchers, their resources and the institutional and organisational contexts (Lemaine et al., 1982). These works describe the intervention of miscellaneous factors in scientific activities, with an often more epistemological than anthropological leaning. They advance no ambitious thesis and their works are not theoretical manifests. In a similar perspective, Terry Shinn (1982, 1988) follows comparative social and epistemological analyses of physics, chemistry and IT laboratories.

In 1975, the French philosopher Bruno Latour, returning from a survey he had carried out in Ivory Coast on the reasoning of African managers, joined the American biochemical laboratory of Guillemin who invited him to carry out an epistemological study. Unfamiliar with the sociology of sciences and with a very poor command of English and never really having taken an interest in the sciences, Latour adopted an ethnographical, 'naive' approach to understanding the conceptual and practical culture of the laboratory. He familiarised himself and worked there as a laboratory assistant. He reports on the inscription games within the laboratory and with the sociologist Steve Woolgar, published *Laboratory Life: the Social Construction of Scientific Facts* (Latour and Woolgar, 1979), a work which is considered a pioneer and one of reference. Hard on the heels of his work and in conjunction with the engineer Michel Callon, who had become a sociologist, the Centre de Sociologie de l'Innovation (CSI) from the Paris Ecole des Mines, developed a theory of the actor-network (ANT) whose influence was to gather pace in the 1980s and 1990s as well as the practices used for field surveys (including Vinck, *Du Laboratoire aux réseaux*, published in 1992). CSI researchers pay attention to the innovation process, the discourse regimes and technical democracy in particular.

At the same time, the German sociologist, Karin Knorr-Cetina, inspired by ethnomethodology, studied a biochemical laboratory in Berkeley. She was interested in the practical informal reasoning of researchers in the working environment and as part of a constructivist perspective. She wrote *The Manufacture of Knowledge* in 1981. Her analyses extended to models of epistemic culture (1999). The work of Merz (1999) on simulation practices in high-energy physics is part of the same perspective.

At the same time, Harold Garfinkel, Eric Livingston, Michael Lynch (Garfinkel et al., 1981) and Steve Woolgar, founders and disciples of ethnomethodology, also took an interest in science laboratories and in the mathematician's office to look at the emergence and practical accomplishments in situ of an order of knowledge. Woolgar wrote *Laboratory Life* with Latour in 1979. Lynch wrote '*Art and Artifact in Laboratory Science: A Study of Shop Work and Shop Talk in a Research Laboratory*' around 1978/79 (published in 1985). Livingston published *The Ethnomethodological Foundations of Mathematics* in 1986.

Other sociologists took up laboratory research as part of a constructivist approach and concentrated on scientific practices such as Law and Williams ('Putting facts together: a study of scientific persuasion', 1982) or Zenzen and Restivo (1982).

Coming out of the sociological tradition of the pragmatic school (John Dewey, George Mead, Arthur Bentley) and the symbolic interactionism school of Chicago (Herbert Blumer, Anselm Strauss, Howard Becker), Joan Fujimura, Susan Leigh Star, Elihu Gerson (the Tremont group) and then Geoff Bowker also took an interest in scientific practices.

Sharon Traweek, an American anthropologist, undertook a vast survey at the beginning of 1980 in the US and Japan on particle physicians. In 1988, she published '*Beamtimes and Lifetimes: The World of High Energy Physicists*'. Her analyses were of the symbolic anthropology ilk. Valéria Hernandez (2001), an Argentinian anthropologist and disciple of Gérald Altabe in Paris, focused on the anthropological foundations, within a laboratory, of demarcation (scientific, non-scientific) and power relationships.

British sociology of sciences from the strong programme in turn paid particular attention to the laboratory surveys and initiated further investigation around the discourse analysis, reflexivity and new literary forms with Malcolm Ashmore, Nigel Gilbert, Michael Mulkay, Trevor Pinch and Steve Woolgar.

The laboratory studies stimulated field studies concerning technological practices (*Everyday Engineering: An Ethnography of Design and Innovation*, Vinck, 2003) or science policy practices (Cambrosio et al., 1990). Laboratory studies are now looking at the nanoscience sector (Fogelberg and Glimell, 2003; Vinck, 2006).

The following pages concentrate on these laboratory studies in order to explain some of the analytical approaches and objects which caught the attention of researchers: hierarchy, language interaction, production of facts, instrumental and literary practice.

The Articulation of Scientific and Social Practices

Shinn (1983) showed that scientific productions are unequally distributed according to the laboratory's hierarchical grades (from the young researcher to the boss). He observed a correlation between the cognitive hierarchy of results (scientific importance relating to types of work: empirical studies/theoretical interpretation) and social hierarchy (the status of members of the laboratory). In this experimental physics laboratory, the researchers work individually on different phenomena. They are in competition for the resources required to construct their equipment. Furthermore, involved in teaching, some develop capacities to produce global, synthetic and detailed representation of phenomena, transmit their personal interpretation and switch from one phenomenon or one model to another.[1] Each researcher fulfils all the tasks connected with it, designs and constructs his or her experimental mechanisms himself or herself as well as the instruments, carries out the experiment and analyses the data. In this context, there is correspondence between the social hierarchy of the laboratory and the type of research results. Shinn distinguishes three groups of researchers and results:

- **The young** produce local results, show the diverse facets of the phenomenon and emphasise their complexity. They avoid overgeneralisation and, on the contrary, pay attention to the conditions of validity and to irregularities. They reject simplification which is in contradiction to the fine tuning of their analyses. They are attentive to instrumentation and its precision. They talk about the precision of measurement and the terms they use. They are sensitive to the multiplicity and relativity of interpretation. Their results are of a compilative, detailed, precise type which are open to criticism. Their local results are only recognised or, indeed, are only of interest to a limited number of other researchers.

- **Senior researchers** put the emphasis on the selection of phenomena representation models and on the integration of data into a carefully chosen model. They compare models and integrate their data into them. They emphasise the conditions of validity and irregularities when they are significant and stimulating for the study of new phenomena. The synthetic interpretations they provide give access to a wider audience.

- **The director of the laboratory** produces more 'generalisations' than anything else, combining several categories of phenomena within a simple and predictive model. He (or she) restricts the number of parameters taken into account and emphasises their key features. He/she will spend little time on the conditions at the limits and on irregularities. His/her argumentation is simple and structured. He/she integrates his/her personal observations outside the laboratory, takes work carried out in other laboratories into account (via information obtained during informal conversations) and refers to the dominant literature of his field. His/her results federate different phenomena and offer a heuristic which allows him/her to structure research activities for his laboratory. His/her analyses help a large number of researchers, orientate their work and structure their research activities. The integration of experimental and cognitive materials from several sources (his/her own laboratory, literature and the many laboratories with which he/she has links) ensures that his/her results gain prestige which local results from other researchers do not.

The social context (hierarchical position and the researcher's social network) and the content of the work (type of scientific investigation) come together to determine the social hierarchy of the laboratory's scientific results (Figure 7.1). However, the social hierarchy of results does not always correspond to the cognitive hierarchy. When they diverge, they are a source of conceptual questioning (challenging of results) and social questioning (challenging of the authority of the director and weakening of the organisational structure of the laboratory).

The Constructed Character of Scientific Productions

Unlike Shinn, where scientific results reflect the irregularities of phenomena and transcend researchers, laboratory studies demonstrate in particular the

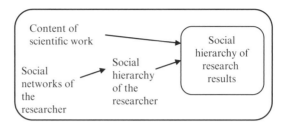

Figure 7.1   *Determining of the social hierarchy of research results*

constructed nature of scientific production (their analysis is qualified as 'decon-struction' (Box 7.2)). The facts appear as the result of the interweaving of a mul-titude of actions and events. The analysis accounts for operations, negotiations, fiddling and tests whose intricate nature will lead to facts, data, concepts, instru-ments, methods, organisations and so on.

As a result, laboratory studies end up multiplying the number of elements (actors, events, instruments and so on) that need to be taken into consideration to report on the resurgence and stabilisation of a scientific fact or statement.

Box 7.2   *The case of TRF (thyrotropin releasing factor)*

Latour and Woolgar (1979) report on TRF by identifying the actors involved, their relationships and strategies as well as the material resources they use (samples, animals), the terminological resources and the instruments at play. The 'TRF' concept corresponds to a heterogeneous network of humans and non-humans whose extension and density of relationships explain its robustness. By noting the multiplicity of entities involved behind the TRF concept, the authors show the consistency, which is constructed through multiple operations and negotiations.

Scientific productions result from a sociotechnical construction in which the distinctions between nature and society, content and context, object of knowl-edge and knowing subjects . . . become complicated. Callon (1986) and Latour (1987) suggest treating these elements symmetrically, whether facts of nature or society, made up by the assembly of materials, people, instruments, calculations and negotiations, and to report on the constructive associations, which are idi-osyncratic, that is, linked to the conditions of their emergence and creation. The scientific constructs have a local and contingent character whose transformations are followed when they are worked on, enriched and reinterpreted by others (Galison, 1987). Scientific practices as they appear in these analyses have nothing to do with the stylised version of writings on science. The activities carried out by researchers are multiple therein: negotiations with equipment suppliers, setting up of manipulations, negotiations on research orientation, interpretation of results produced by the instruments, and the writing of articles. They contribute to a collective adventure in which material elements occupy a substantial place.

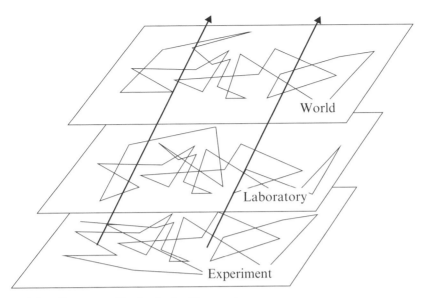

Figure 7.2   *Alignment of levels of scientific practice*

The Construction of the Problem: An Articulation Work

Problem-solving is not automatic. Observation of research practices in laboratories shows that researchers dedicated much energy to constructing the problem
and constructing it in such a way that it is 'do-able' according to the resources
that they can access within and outside the laboratory (Fujimura, 1987). They
undertake a work of alignment and articulation of resources in terms of the
experiment and the manipulation to be carried out (organisation of the space and
the sequence of work), their integration into the life of the laboratory (negotiation of an access point to an instrument requested) and in the world (ensuring the
availability of a sample which a hospital or industry is to provide them with). The
work involves associating everything which seems necessary for the proper execution of the project. It is made up of planning, organisation, control, evaluation,
negotiation, adjustment and integration activities, at the level of the experience
(a set of tasks), of the laboratory (a set of experiences and coordination tasks, in
particular the management of instruments) and of the world (made up of laboratories, colleagues, funding bodies and others). This work involves exchanges
with other people (to obtain authorisation to order a specific reagent, to ensure
the availability of a technician for a delicate instrumental operation). It involves
interfacing the tasks of everybody at these levels and aligning the different levels
in relation to each other (Figure 7.2).

   As a result, the definition of the research topic depends less on the social
group or its social interest than on the ability to mobilise resources in order to
carry out the experiment – use the animal model developed within the laboratory,
more available than a model seen in the literature – and coordinate it with other
researchers – there can be competition in the use of certain instruments.

The content and consistency of scientific work depends on elements which researchers articulate, elements whose origin and trajectory influence the work in progress.

The situations in which researchers have to face up to unforeseen circumstances or errors which affect the course of their work (for example, the non-availability of a piece of equipment and the use of an alternative solution), are particularly interesting to study. In these cases they have to stop the ordinary course of action, ask further questions and carry out critical reflection which is not required as long as everything goes as planned. They are, therefore, required to formulate and test hypotheses on this interruption, try out new arrangements, redefine pathways and new organisations.

The researchers dedicate time to constructing aggregates which facilitate the articulation work: protocols (interfacing tasks, instruments, products and human resources), instruments (interfacing operational programmes), activity reports (interfacing the activities of the laboratory and giving them sense), partners' clubs (interfacing financiers, colleagues and users who support the activity). The aggregates transform into a whole, a multitude of singular elements. Thus, they aggregate groups of tasks and set up 'black boxes' which can be used without having to reconstruct the internal interfacing. The production of aggregates like this, such as that of standardised interfaces, facilitates the interfacing of social universes and produces accelerator effects, hence the importance of the work of standardisation of instruments, equipment, methods and concepts to which researchers commit time. Vinck (1992, 1999) shows equivalence phenomena with the notion of 'intermediary object', in the case of scientific cooperation networks and in design process. Star and Griesemer (1989), analysing the work of the cooperation between groups of researchers from different disciplines, also show the key role that the defining of *boundary objects* can play.

An implicit project-based organisation

The activities of the laboratory are often structured by projects, not in the formal sense of project management methodology inspired from the industrial world, but in the sense of 'sequential unities', the result of which is the authoring of a research report or a publication. The project seems to be the organisational unit which allows us to allocate tasks to members of the laboratory, to order supplies, to prepare equipment, to propose phenomena to be studied and to orientate bibliographical research. The projects are linked between them, but their links are multiple and complex. Also, they are contingent adventures whose result can never be 100 per cent sure. Their continuity does not result from their initial planning. On the contrary, they are always likely to be interrupted or abandoned, reorientated, transformed, differentiated or merged.

Projects are not visible as such when we visit a laboratory. They do not correspond to sequences of tasks with clear spatial and temporal interfaces. On the contrary, tasks are carried out simultaneously by different individuals and it is not easy to identify either interfaces between these people, or links between different tasks carried out by a single person. For example, it is difficult to know if two tasks executed successively by the same person are to do with the same project or to do with different projects. Researchers and technicians organise themselves to be able to carry out tasks concerning several projects. Waiting times (during the incubation of a cell culture, centrifuging or the time when the results are in the hands of the boss) are used to carry out tasks concerning other projects. A single task can also be involved in several projects simultaneously (for example, the preparation of experimental material). Finally, certain tasks are interrupted or deferred in order to switch to another project, for example when repeated difficulties are met. In this case, the unfolding of the project is suspended in order to move on to a different enquiry or a new project (for example, the design and preparation of an instrument) whose result will enable work to continue on the initial project. Such enquiries can overlap and lead the researcher on a long detour before s/he returns to the initial project. Sometimes they lead the researcher to other very different projects from the one initially planned.

The course of action, contingent and nevertheless decisive

The invisibility of projects is such that neither the outside observer nor members of the laboratory can have an overview of the activities going on there. The research methods and protocols do not report on the actual sequence of activities. The problem is not due to the fact that they are improperly designed or insufficiently detailed, but to the practicalities of action. An experimental protocol addresses a single procedure (for example, fixing a vascular profusion to rats) for the different cases treated (the same procedure for all rats). And yet, in practice each case is different (each rat reacts differently or the researcher's actions are not perfectly constant). Each time the procedure is implemented in a specific manner.[2] It follows that the comparison of results is always something of an issue.

The series of effective actions is much more complex than the methodological description because the method prescribed is based on tacit competences which are supposed to be part of a shared practice approach. It also implements a series of repair operations in order to cater for unforeseen events. Thus, there are many ways of linking the capricious nature of a product or instrument. In this respect, Lynch talks about 'superstitions' and 'personal preferences' of researchers linked to certain procedures. Finally, although the tasks of a project can be carried out by different people, the work is often carried out by a single person. The reason given is the need to have access to the background of the procedure through which the phenomenon has been made visible. This point of work organisation is sometimes the subject of high controversy in institutions where the rationalisation of research activity involves the setting up of a platform and the introduction of a demarcation between researchers who interpret the operations to be carried out and the technicians who actually do them (Vinck, 2006).

Mobilising Resources: A Reconfiguration Exercise

In the constructivist perspective, the laboratory is not a physical space within which the nature, cognitive factors and social factors are superimposed. It is a mechanism where heterogeneous elements are interfaced (resulting from different trajectories) modelled by work groups and by material operations.

The elements which are mobilised and articulated within laboratories are rarely the subject of nature, but rather traces or formatted samples of purified and transformed versions of nature. The laboratory moves objects taken from nature within its walls. In the same way, it is not concerned with events which occur outside the laboratory's intervention or control. With the laboratory, on the contrary, we see the researcher impose another temporality on events. S/he detaches objects from their natural environment (makes them mobile and malleable) in order to be able to introduce them into a new socially constructed phenomenal field (it socialises objects from nature). S/he makes a culture of objects from nature within the laboratory's sociotechnical mechanism.

In the same way, the laboratory installs and reconfigures scientists by making them operational and adaptable. They are part of the laboratory equipment and its research strategy. They are the depositaries of a capacity to make sense of nature and vehicles of a subconscious experiment which can be mobilised in order to solve problems. These competences, which are partly tacit and incorporated, are only partially to do with the conscious activity. Researchers function partly like instruments; they are laboratory resources, formatted entities which are there in order to interface effectively with other entities.

These elements (humans, equipment, concepts and so on) are transformed during the time they spend in the laboratory. They are made malleable and contribute to making the laboratory an area for new phenomena, a new phenomenal field and an instrument for the reconfiguration of natural, technical and social orders. By instrumentalising researchers and socialising nature, they become a mechanism from which a new order emerges which is neither natural nor social (Knorr-Cetina, 1981, 1995). It produces a 'socionature', that is, a world which is both natural and social.

The Make-up of Phenomena: A Contingent Exercise

Analyses of laboratory practices show that the phenomena studied are not so much revealed by experiments as produced by contingent details of the practice and adjustments. The visibility and appearance of phenomena and of objects are constituted (and not reflected) by the instruments and by practices used. In the classic conception (reflexive), instruments reveal phenomena; their influence on the constitution of facts and their visible appearance are ignored.

The constitutive role of practice appears when an artefact emerges (unexpected event which is attributed to the practice or to the instrument). In these situations, the researchers put the results down to their action and not to nature. They express their surprise ('Oops', 'What's that?' or 'There's something wrong here'). Their attention focuses on the history of circumstantial events which have led to this artefact. They suspect the procedure, the equipment or the instrument. They focus on the sociotechnical mechanism and reveal its role in the constitution of phenomena. The artefact manifests the failure of the habitual auto-effacement of laboratory practices. The 'natural phenomenon' appears, therefore, to be the result of a sociotechnical action which is usually ignored.

The study of what occurs when an artefact emerges reveals a lot about the role of practices in the constitution of phenomena. The artefact, more than the expected natural phenomenon, emerges during the action, as a problem (an incongruous presence or absence), as the intrusion of an object hidden in the production of the visibility of things. In this situation, researchers will often start talking and discussing. They complain about the material used (sample, reagent or instrument) and consider the artefact as a result of an error or a problem associated with the procedure:

- Certain artefacts are **linked to the standardisation of laboratory work** (instrumentation, preparation of equipment). Researchers present them, therefore, as examples of an improper manipulation.

- Other artefacts are **linked to the novelty of the technique**. In this case, the very fact of being an artefact is treated as a problem and can be the origin of controversy as to understanding whether it is a natural phenomenon or an artefact. Once the artefact has been localised and linked to a cause, it can be isolated, which can lead to this equipment no longer being used.

- Sometimes, the material is removed from the process because of its **lack of 'beauty'** and the risk of being accused by readers of doing poor-quality work and producing ambiguous evidence. For public presentations, therefore, they will keep only 'good' results.

Certain artefacts are considered *positive* to the extent that they can be linked to sources (via a genealogy of the artefact). Other artefacts considered *negative* are not linked to any technical factor which can be reported on. The members of the laboratory will then talk about procedures (capricious or otherwise) that they cannot explain. Ensuring that it works is a fragile process whose solutions are not pre-ordained, especially when the researchers are never sure of the existence of the presence of the event or the object that they are trying to study. When an experiment fails, they have questions on their hands: is the phenomenon absent or has the experiment been done badly? Is there anything that we could do to make this work? These questions remind us that the results are contingent, that they depend upon competent actions and that there is a substantial link between the things represented and the technical procedures.

The Creation of Agreement or Objection: A Negotiation Exercise

The 'fact' rarely imposes itself. Researchers learn how to produce it and how to distinguish it from the artefact thanks to various manipulations and critical examination. They talk about it and agree on what they see on the facticity of the phenomenon or the reliability of the result. The agreement on fact comes neither from the evidence that nature imposes nor from a tacit agreement or pre-existing paradigm which leads researchers to think in the same way.

On the contrary, agreement is based on interaction. It results from situated, empiric procedures (Lynch, 1985) which it is down to the sociologist to study. In this way, studies show that: (i) agreement is affirmed by language ('yes', 'ok') or by some sort of gesture. Affirmation is orientated towards a statement or a gesture which comes before it; (ii) it is imminent to the situation; (iii) it is local; it concerns those who are involved in the situation and is close to another enunciation to which it refers; (iv) it is independent of the fact that the parties concerned are or are not in agreement. It can be affirmed without the enunciator really believing what s/he is stating. The agreement is a factual event to which the parties can come back so as to identify more closely what they expressed as agreement; and (v) it is not anything other than its production or local recognition. It is part of the action and demonstrates its collaborative dimension.

The production of the agreement contributes to the production of results which will then be publicly affirmed (Figure 7.3). It reinforces actions in progress (hesitant expression of a hypothesis, assessment of a manifestation as a fact or artefact) whose outcome is uncertain, but determined by sequences of agreements and disagreements (Box 7.3) which are empirically observable during which the announcements are modified depending on marks of agreement and disagreement expressed by colleagues. Objectivity results from the agreement established between the members involved in the situation (ibid.).

When agreement is reached, that is, when there is no sign of disagreement, the language interaction ceases or changes subject. The agreement operates a ratification of the description while the disagreement prolongs the investigatory and modification work until the new affirmation can no longer be called into question. Analysis of these language exchanges shows how scientific discovery (ratified descriptions of objects) are shaped by social interaction. Once agreement has been reached, only the result is taken into account; the object is considered to be obvious. Its social fabric disappears into the shadows. The linguist Mondada (Mondada and Racine, 1999; Mondada, 2005) reports on the emergence of the construction of concepts, scientific affirmations and technical norms as collective achievements. She shows how objects of the discourse are proposed, reworked, ratified, transformed or rejected by people exchanging views and how they construct a collective version of the description of the world they are studying.

Andrew	I've six …
Luce	Sure ?
Andrew	believe this one is an artefact ?
Luce	Mmh …
Andrew	And these ?
(silence 4 seconds)	
Andrew	Then … we have two

Figure 7.3 *Sequence of production of an agreement*

Box 7.3   *A conversational sequence*

A researcher states something concerning an object. The tone shows certainty and the definitive character of the affirmation. A colleague defies this affirmation with a counter affirmation, a silence, a 'Really?' type question, a sound ('Hmm') or a gesture. There follows reaffirmation which is modified by the first protagonist. Something of the first description is maintained. Sometimes the first affirmation is repeated more loudly or with a less affirmative tone. Reaffirmation can be defined once more, provoking further modifications. Among the observable modification throughout the interaction, we can note the following:

- **the redefinition of the scope of the reference.** With a term such as 'the same' (which can be understood as 'exactly the same' or as 'more or less the same'), the scope of the reference can be modulated as the conversation progresses. Its meaning is circumstantial;

- **the relativity** of the affirmation through the addition of expressions such as 'I think', 'I suppose', 'I don't know but . . .'. The first affirmation had no author (the announcement was simply affirmed), while in the reaffirmation, the author clearly states him/herself as the source of the announcement. S/he, therefore, leaves a doubt and shows him/her sensitivity to the disagreement expressed by the other party;

- **the addition of explanations**. While reaffirming the first statement, taking the disagreement element into account sometimes leads to the revelation of new elements, nuances, more precise explanations or details which initially were not in evidence. The new description is supposed to be able to resist the reaction of colleagues better.

The Researcher's Word: A Made-to-measure Exercise in Conviction

Laboratory studies show an enormous contrast between science as it is discussed in the laboratory (talking science) and science about which we speak (talking about science) (Box 7.4) where very little attention is paid to conversations at work which seem to be either without interest or too difficult to understand.

Box 7.4   *Laboratory visits and 'demos'*

These means are used by researchers to show and talk about the work to people coming from other laboratories or from industry, to potential students or observers such as a sociologist who may want to see things with his/her own eyes.

Demos (Rosental, 2008) have an ostentatious orientation in relation to work in progress and are an opportunity to exhibit phenomena, operational principles and to illustrate the value of a given approach. Their preparation and implantation are structuring elements of the scientific activity. They hide as much as they fulfil a test function with regard to the next stage reserved for research projects. They are a form of presentation of self for researchers as well as a way of meeting others, a means of prospecting and a form of gratification with regard to a visitor.

Laboratory visits are conducted by members of the laboratory in such a way that they hardly interrupt the activity going on there at all. The guide or his/her colleagues talk about what they are doing, show slides, materials or results, give advice, tell stories and answer visitors' questions. What they say will be formulated according to the type of visitor. When the visits are organised for colleagues, they include numerous question/answer sequences in the form of an information search or, more often, in the form of a challenge made by the visitor to a member of the laboratory. These challenges concern controversial or uncertain points. The result is that laboratory visits take on a form of defence of projects faced with visitors who consider themselves to be competent enough to understand the work being done locally.

Visits seem to show science as it is being done. However, when observers extend their stay, they discover that it is not quite what it seems. When researchers are concentrated on their work, you can no longer question them about what they are doing and why they are doing it. They are impatient and sometimes ask not to be disturbed until the operation is complete. Effective laboratory work does not lend itself very well to direct oral account. In contrast, laboratory visits and demos seem to be staged events and 'representations' of the work being done.

In practice, researchers explain very little of what they are doing. When there are a number of researchers working together, it looks as if some invisible force is ensuring agreement between them. The language used is not descriptive; it is part of the action and the work process: reaching an agreement or a disagreement, asking, suggesting, declaring, assessing, doubting, drawing attention to and so on. Also the language acts of the scientist are not limited to spoken language; they include ordinary non-lingual sound expression (for example, a vast range of: aah, a-a-a-ah, oh, oooh, hm, mmh, yes, y-y-ess) and gestural and graphical communication. Phrases are also interspersed with periods of silence whose duration and intensity are significant.

These language exchanges are, however, a lead in to more constructed, structured and rational discourses which may be produced. They allow researchers to come up with ideas and test them and to elicit reactions without investing too much in the formulation of well-constructed statements.

For the observer, these exchanges are not easily accessible. It is not sufficient to observe them to understand them, even with the use of audio-visual recording. They are opaque because of their highly technical nature, their implicit reference to what has already gone on in the past and elsewhere or their reference to what is in progress. Understanding them is inseparable from the situation; they are embedded ('indexicality'). They are phenomenologically strange to the naive observer while resembling practically ordinary conversations. The opacity of laboratory conversations is not to do with the technical vocabulary used, which is relatively absent, but to the unusual use of ordinary terms, pronominal forms and 'pro-verbs' (such as 'do', 'work') and 'pro-nouns' ('the thing', 'the animal') which refer to the situation and to the circumstances of which the speakers are supposed to be aware. The same goes for silences which are sometimes replaced by a manual operation which overlaps the conversation.

These characteristics are distinguishable from other language-based or written scientific productions which are, in contrast and in principle, de-contextualised, transparent although esoteric.

The Production of Scientific Statements: An Exercise in Inscription and Modalisation

When Latour and Woolgar (1979 [1986]) describe the laboratory, a simple principle emerges and gives sense to all the activity going on there: the office of the researcher, reader and author is the pivot of the laboratory. Towards that office converges literature coming from outside the laboratory, as well as papers produced within it. On his/her office, the researcher juxtaposes these two forms of literature, annotates them and establishes cross-references between them. Some of the documents on which the researcher's work is based come from the laboratory where the instruments are located and where s/he observes an intense activity of registration: interminable lists of figures, marking of test tubes, labelling of and even writing on rats. The presence of writing and their collection formats (mail, invoice book, lists of data, photocopies of articles, library) lead us to conclude that the tribe inhabiting the laboratory is characterised by an intense activity of coding, marking and writing.

Laboratory manipulations result in the production of written documentation, data and graphs. The instruments are inscription devices; they force the study objects to express and produce a sign of their presence. The written evidence produced, therefore, takes on more importance than the products and the instruments. Transformed into graphs, they catch the attention until they are incorporated into a text. The laboratory activity can then be resumed into the production of a graph which accompanies a text. According to Latour, the scientific activity could thus be explained without having to have recourse to epistemological concepts. The laboratory is a mechanism for literary inscription[3] and the production of images (Latour, 1986; Galison, 1997). Study of the preparation of visualisation instruments, methods of sample preparation and processing of evidence shows how scientific objects are brought into existence. Medical X-ray images will only show us something after the techniques have been fixed and after a certain way of preparing the patient, but also as a result of technology to isolate what there is to see (Pasveer, 1989). Other works show how meanings are collectively constructed on the basis of visual clues (Lynch, 1985; Knorr-Cetina and Amann, 1990); how the elements are selected and qualified as significant as opposed to being considered as basic noise; how a purified image emerges; how images from experiments and simulations are matched up; how the 'good' shots which could be considered 'showable' are selected and how the shots are schematised and mathematicised.

An exercise in (re)$^{n-}$representation

The objects the scientists are working on are transformed through a series of operations which lead to representations being constructed. At the beginning the phenomenon is represented by a few samples which, once processed, are converted into graphical representations (traces, figures, codes), and then into diagrams, models and texts. Animals taken out of nature are moved and trans-formed to become zoological specimens via the traps laid for them, the labels that are placed on them, and the suitable preparation for their transportation and conservation. From being dispersed in nature, they are brought together and lined up and then transformed into words (labels, datasheets) and into points on geographical maps of fauna.

Each stage uses a prior representation of the phenomenon as a point of departure. The 'final' representation results from changes in series which consti-tute 'chains of (re)$^{n-}$ representation (Latour, 1995). Each stage removes certain aspects of the original phenomenon and emphasises or transforms others. Interactionist sociologists, analysing the construction of these representations are looking at the perspectives of actors which influence them and their reception by those at the next stage who make an order representation of them ($n + 1$). Given the work of simplification which leads to removing part of the prior inscription work, differences of interpretation can emerge between those who produce a representation and those who use it. Then there can be shifts in representation. Fujimura (1987) uses the notion of 'problem path' to trace the changes in struc-ture of problems and resources and 'bandwagon effects' which transmit from one version (n) of the representation to other versions ($n + x$). The study of routines and non-problematic paths is completed by the study of problematic situations, suspensions of routine, irregularities and accidents which lead to a search for repair. Analysis describes the route via which the event is qualified as abnormal (entered, delimited, isolated, made visible, categorised), linked to other irregulari-ties, to categories of irregularities and other events. The confrontation of several situations of this type allows us to detach from the interpretation that actors give while using these interpretations as a basis.

The inversion of the relationship between inscription and the object of nature
When a fact is accepted as being scientific, traces of the context of its produc-tion disappear. The authoring processes thus lead to publications which give the impression that the facts speak for themselves. This effect is the result of a process of dissociation and the inversion of the relationship between nature and its representation (Woolgar, 1988):

- The scientist has traces and inscriptions, from the instruments s/he uses or from literature: (inscriptions).

- On the basis of these inscriptions s/he operates comparisons and combinations in such a way as to show an object. It gives form and existence and constitutes his/her object: (inscriptions → object).

- It represents his/her object as something which would be independent of these constitutive inscriptions. It dissociates the object from the inscriptions from which it emerges: (inscriptions/object).

- Thus, it reverses the relationships between the object and the inscriptions. While the object emerges from inscriptions, it acts as if the inscriptions were the reflection of the object: (inscriptions ← object).

- Finally, in order to support this inversion, the three first stages are minimised or, indeed, even forgotten.

The work of the laboratory involves producing inscriptions and formalising factual statements in such a way as to liberate them from the local circumstances of their production.

The modalisation of statements

The process of inversion and deletion of the work of producing objects involves the authoring of publications where 'scientific facts' appear in the form of literary statements whose formulation itself is negotiated. Certain statements are loaded with numerous references (to other texts, instruments, equipment and the method used) which contribute to persuading the reader. The other statements which concern things of which the reader is already convinced, maintain no trace of their local reference universe. Thus, a relationship between the degree of facticity accredited to a fact and the literary genre in which it is announced emerges. Facticity varies depending on the modalities of the statement (Box 7.5).

Collective scientific dynamics and work thus result in the production of factual statements, switching them from a status of opinion of an individual to a status of fact established and recognised by all. This work involves writing and modalising the statement. Peers often test these statements and attempt to retrograde them to a status of opinion, hypothesis or artefact by adding nuances or doubts. Some see their modalities oscillating between confirmation and denial while other modalities are instituted as 'established facts', and are included in manuals, incorporated by socialisation (tacit knowledge) or materialised in instruments.

Latour's sociotechnical and semantic networks

Latour and Woolgar (1979 [1986]) thus describe the process of collective construction of the chemical structure of a hormonal releasing factor of the brain. The statement of the structure is stabilised by microsocial interactions, the production of inscriptions, their superimposition and transformation into modalised statements depending on discussions and criticisms received. The statements received are transformed progressively from their path of one author to the other. Sociological analysis must then describe the networks of people, texts, terms, materials, instruments which from neighbour to neighbour become associated and interfaced. The reality of the fact or the sturdiness of a statement depends on the extension and knowledge of sociotechnical and sociosemantic networks which are constructed collectively. Each element of

---

Box 7.5   *Modalisation*

The statement A–B, is a factual statement stating a relationship between A and B, for example:

Task X is nebula. (1)

This statement, once slightly modified, can become:

Dupont suggests that task X is nebula. (2)
Dupont says that task X is nebula *because he does not want to bring into doubt the quality of his observation.* (3)

In modalisation (2), the statement introduces an author (Dupont) and his action (he suggests that). The initial statement (1) was supposed to be independent of any author. In the statement (3), motivations are even accredited to the author. Depending on modalities, the statement switches from the status of 'scientific fact' to one of 'personal opinion' or vice versa. If the factual statement convinces, the modalities disappear; if it is not convincing, modalities are added, for example with the addition of local circumstances. Modalities can affect the relationship (4) or one of its terms only (5). Thus:

Task X *is supposed to be* nebula (4)
Task X is a *'nebula'* (5).

The author of a statement can therefore increase (or reduce) the facticity through elimination (or the addition) of references to an agent (the researcher, the author of a text, an instrument), to his/her action (s/he affirms, challenges, supposes, demonstrates, resists) and the circumstances of this action (motivations, contingencies). When s/he introduces elements concerning the interest of an actor or the circumstances which explain his/her gesture, s/he produces a statement which concerns both nature and society. *S/he defines actors and establishes relationships.* Conversely, without modality, the statement creates the impression that society has nothing to do with the fact.

---

the network is itself linked to a trajectory via which other associations appear. The scientific work involves constructing, extending and stabilising such networks. This means constructing and stabilising each relationship: making two instruments compatible, harmonising vocabulary, increasing exchanges between laboratories, understanding the workings of an instrument, reproducing experiments, articulating statements, or signing a research contract. The series of transformations, translations, relations and travel reveal what emerges from the action.

Scientific Writing: An Exercise in Persuasion

Researchers elevate their activities to the level of discourse, in particular through the construction of texts. We shall now look at these writing and signature strategies.

Scientific writing

Writing strategies vary according to the type of publication: an article for a specialised scientific journal, a poster submitted for a conference, a handbook, patents, research reports, research proposals, annual report, mail and training manuals. Scientific literary production is substantial. It is a major source of information in many studies into sciences; scientometry develops methods of quantitative analysis on this base. Among the articles, several literary genres are noted:

- Articles destined for a non-expert public contain **general terms which may catch the attention of readers**: the benefits of science, potential applications, national challenges. Contextual details and the production of knowledge are absent in these texts.

- Articles targeting a scientific audience working in other domains. A more specialised vocabulary is mobilised so as to **link up the author's area to questions and objects of colleagues** who will potentially use this knowledge: researchers, decision makers, doctors or engineers.

- The articles which are addressed to specialists from the same research field in order **to bring knowledge up to date.** These are syntheses, state of the art and reviews of challenges to be met and questions which remain unanswered. They are often signed by several different authors.

- Articles which are addressed to specialists from the same research field in order to communicate new elements of information. Their titles are **esoteric** and the text is loaded with references to other texts, to data lists and to graphs.

The writing of scientific texts is a practical exercise which can be analysed in the same way as other laboratory practices; these are more or less collective constructions which are dotted with returns to the drawing board, mixing and juxtaposing inscriptions from instruments and from the library. They are the subject of discussions, crossings out and corrections.

Writing is not equally distributed between all actors. Technicians who are often consulted during the writing are not called upon to write themselves. Some researchers who are reputed to be good writers in the laboratory will often be made responsible for the writing phases while the director of the laboratory will exercise control over those writings and even do the re-writes of the final version in certain cases.

Writing is also relative in the sense that the same event is presented differently according to the public targeting (Woolgar, 1980; Law and Williams, 1982; Latour and Fabbri, 2000). Statements undergo significant transformations through the writing process. Writing is rhetorical: the objectivity is formulated through the use of particular syntax, the choice of terms and socially imposed or authoritative terms. The style and literary technologies (Box 7.6) of codification of the reports on experiments (Shapin and Schaffer, 1985) have indeed evolved substantially (Licoppe, 1995). Modes of writing, demonstrative and explicit

Box 7.6   *Textual mechanisms which are identifiable within publications*

Preliminary indications: the formatting of text and the academic journal where it is to be published are indications given to the reader which lead him/her to perceive it as an authoritative piece of writing and not fiction. The mention of the institution to which the author is attached and the bodies who support his/her work suggest that s/he is not speaking alone; behind him/her there is a network which s/he presents through the text. Through the words in the title and the key words of the text, the reader is informed of the fact that the text refers to entities which are supposed to exist independently of the text. The summary suggests a problematic situation or tension ('until now, we knew . . . but . . .') and a solution. The solution is constructed in the text as if something pre-exists the text and the research and as if the text only shows how we got to this point.

Externalisation: the text presents the phenomenon as if it has an existence which is independent of the text. It produces the impression of the author's non-involvement as s/he 'writes under the diktat of nature'. The phenomenon is presented as something which goes beyond the text (out-thereness). The author uses the passive form and writes as if the researcher has nothing to do with it. No personal name appears within the body of the text; in the same way, personal pronouns are not used at all. Such a rhetorical procedure creates the impression that nature is speaking for itself, that nobody is speaking on its behalf. The social dynamic of research is cast into the shadows. The use of the passive form (and Licoppe shows how this became the norm) reinforces the impression of absence of the author from the action: 'the result suggests that . . .', 'the fact X leads to . . .'. The scientific actor is presented as passive even when s/he is presenting his/her work (Extract no. 2). Extract no. 1 is that of a novice researcher who has not yet integrated the norms of academic writing:

> Extract no. 1: 'I grew the culture for 2 days more or less, the best strains of A had been given to me by my colleague X. Then I took out several cells according to the method set out in X's manual, but arranged by our technician. I took 10 to be sure to have at least 3 good ones . . .'

> Extract no. 2: 'After 2 days of culture, 3 cells were extracted from the A strains according to method Y . . .'.

The intervention of the author is minimised and creates the impression that any scientist in the same situation will have come to the same conclusions. Details, unforeseen circumstances, hesitations and local variations affecting the experiment are absent. The report, for example, mentions that 25 animals were treated without indicating that the same procedure was not applied to two separate animals. It has not explained how things took place on a case-by-case basis. The series of cases is treated in the text as a set of equivalent events ($n = 25$).

Reconstruction of the history: In a few lines, the author reminds us of the state of the issue and writes *a* history, emphasising continuity ('since X, we have made substantial decisive progress . . .') or change ('in spite of the works of X, it is only very recently that . . .'). Through the choice of authors evoked, s/he decides on the history of the subject. The past that s/he lays down defines the problem as it is posed and invites the reader to follow the movement of history (which is going to be explained through this text and which could be prolonged if the reader deigns to cite the text).

Capturing the reader's attention: in the text, the author constructs an identity for the reader by imagining his/her objections and attempting to respond to his/her expectations and interests (via technical details in the 'material and methods' section).

structures, problematic formulations as well as the nature of entities brought into play have also changed over time.

A scientific publication is a literary genre all of its own. It presents a contribution to a particular readership which seems original and thereby attempts to modify the behaviour of its readers: catch their attention, change their perception, engage them to use the results published and to cite the author.

Since writing is strategic for researchers, the choice of journal in which the article is to be published, references to be cited, details concerning methodology, terms used, publishable data, and the style of tables and graphs are all potentially subjects of debate within laboratories. There is a question of readers, editors, the way of introducing texts, the title, the style (moderate or ambitious), the degree of generality to be given to results and to interpretations (depending on the scepticism or enthusiasm that authors imagine to exist on the reader's side). During writing, they explore their own interests and those of their colleagues so as to discern the best strategies to use.

On top of these laboratory discussions, there are those that occur on editorial panels and the to-ing and fro-ing of negotiations with commentators and then those of readers who use, forget or transform the arguments used. Over time, the text undergoes proofreading which leads to it dissociating its 'textuality' from its 'cognitive content' and incorporating the latter into a body of text through which the value of truth of the original text is redefined (invalidated or strengthened and exemplarised (Box 7.8)).

The formatting of publications is also the subject of negotiation between researchers and with reviews. The size of the article, the way it is cut up, the use of diagrams, tables and photos as well as the stylistic effects used on the characters (size, bold, italics and so on) are defined both by editorial standards of the publication (that produce style formats) and by the habits of the discipline (for example, the IMRAD plan: introduction, material and methods, results, analysis and discussion). For authors, application of these norms is not straightforward; they are subject to interpretation, adjustments and resistance, submission, modification or arrangement. Draft articles often go through several versions before the content is stabilised and formatted.

Signatures on publications
Since research is a collective activity and the research accounts are signed, there is a question as to understanding who signs and how this is decided. The problem is even more important because it is at the heart of cooperative practices and

Box 7.7   *Reading a scientific publication*

1 **The format**: journal, readership, language, style, subjects addressed and place occupied in the journal.

2 **Lexicon** (vocabulary used): what s/he is talking about, the problem addressed, the purpose of the text. How the object is defined and is situated in relation to other objects. Its reference framework (the way it is seen).

3 **Scope of reference**:

   • texts and other authors cited (the text is an inter-text) and the use that is made of these citations;
   • authors introduced through the article (signature, body of the text). Implication or ignoring of authors in the text: through what procedures?
   • anticipated readership (expectations and objections);
   • field of research to which the text refers (laboratory experiments, observation in nature, onsite survey, work on literature). What are the actants present (entities to which the text attributes an action, whether they are human or not), how is the problem articulated (venue, context)?
   • method of investigation used. Relationship between the empirical situation (field, experimentation) in relation to the theoretical argument put forward;
   • concepts and theories used for the purposes of demonstration.

4 The **style** (exploratory and digressive, factual, narrative and so on) and effects it produces on the reader (an effect of objectiveness, empathy, irony, case-making). Use of conjugation, articulations and metaphors:

   • content of the demonstration (explicative variables and explained variables), method of argumentation, hypotheses, exposure approach (deductive, inductive and so on);
   • proportion occupied by the different parts;
   • type of trial engaged, type of proof advanced. The use made of footers, images, graphs, tables, figures, equations and other systems of notation (chemical, linguistic, for example).

5 The **intention** or its completion: what are the conclusions of the author and the points s/he is making?

One way of writing the text involves making modifications (changing the sense of an articulation, substituting one actant for another) and observing the effect that these modifications produce.

All these questions will be better elucidated if the text can be followed during its preparation and the use that is made of it by the readers: see Ashmore et al. (1995).

> **Box 7.8**   *A referential intertext with a systematic probationary pretension (Berthelot, 2003)*
>
> 'Intertext' because it refers to other texts; 'referential' through its ambition to address reality; 'with a systematic probationary pretension' because it is about resisting invalidation from the scientific community.
>
> It is characterised by: (i) an explicit intention for knowledge for the author; (ii) a contribution of knowledge recognised by the scientific community; and (iii) its appearance in a scientific publication area. By combining these three criteria, Berthelot speaks about the differentiation between several categories of text and, in particular, the erudite text which is subsequently rejected by science further to the evolution of standards of scientificity; the scientific text whose results are subsequently invalidated; the pre-scientific text subsequently reinstated in science's patrimonial domain; the illusion of scientific text.

the professional assessment of researchers. Pontille (2003) thus questions the practices and mechanisms about signatures: alphabetical order, decreasing order of importance or depending on level of contribution (the first signatory is the one who has done the work; the last is the senior researcher responsible for the project). Conventions and mechanisms which dictate signature practices vary according to disciplines and reviews. They are subject to negotiation between researchers during which the nature of the scientific work, the notion of author, the dividing line between those who sign and the others (informers, technicians, participants in the seminar which provided the ideas and so on) are decided, along with the contributions and responsibilities of each (in particular in the validation of results). The signature, far from being reduced to a graphical apposition of the name of an author, is to do with the local presentation of research actors while taking into account the standards and requirements imposed by publication, the research assessment mechanisms, the prestige of journals and impact factors.

Discursive practices

Scientists are also orators, during visits to laboratories, teaching, conferences that they deliver and their appearance in the media, before the courts and in political arenas in the event of technological controversies. Following the variations of discourse depending on the body where they are asked to express themselves allows us to understand language, social and political competencies which are part of their practices.[4]

The scientist is also in fact addressing society. His/her message produces social effects (effect of cognitive authority in particular), for example when s/he mobilises the notion of truth or terms such as: discovery, fact, empirical evidence, concept, theory, novelty, error. The notion of truth, according to Bloor, fulfils three functions:

- **Discrimination**: this contributes to the categorisation of beliefs into two categories, true and false.

- **Rhetoric**: this allows us to base a statement through argumentation and contribute to giving authority to the speaker.

- **Materialist conviction**: this allows us to satisfy the need to believe in the existence of a stable world to which such theories refer.

Material and Cognitive Cultures

We have just exposed the different types of constitutive practices of scientific research and its productions. A final aspect of analytical input of the constructivist school concerns the material and cognitive culture that laboratory[5] or historical[6] studies attempt to defend.

Material and Instrumental Culture

Instruments, fact producers and inscriptions produce what can then be treated as an objective phenomenon. The existence of scientific facts depends on these instruments and the way they are used. Material culture (Box 7.9), however, goes beyond the simple list of instruments. It also concerns their integration into the appropriate infrastructures, into work bodies having incorporated the ways of doing things and into local networks connecting up other instruments.

Next to physics instruments, sociology looks at intellectual technologies: calculation methods, metrology, graphic representations and so on. Bowker and Star (1999) thus analyse construction and the role of categorisation (invisible infrastructures). In the same vein, works in progress are looking at what is happening in research laboratories with management technologies, quality management, risk assessment, management and knowledge, financial management, merger of laboratories, electronic messaging and so on.

Research practice also depends upon elementary utilities and technologies: good quality water, gas, filtered air and so on. The disconnection of laboratories from these utilities would lead to the disappearance of practically all the facts studied. These technologies are not always specific, but their interfacing with a set of incorporated instruments and know-how (Box 7.10) makes up a local, material and often literary culture which is often inseparable from other elements in that culture. They are linked by 'embedding'.

The philosopher Ian Hacking (1983) also looks at these issue and emphasises the importance of the intervention of the experimenter; manipulating, developing, calibrating, eliminating artefacts and finding artifices. He concludes that the validity of data coming from experiments is less to do with its theoretical foundation than with the multiple interventions which allow us to see and connect what is demonstrated by one instrument with what is demonstrated by another. Their

Box 7.9   *The case of nanosciences*

Researchers pay great attention to these instruments which condition access to the subject of their studies; the fabrication of objects and phenomena studied (of nanometric size), control of their manipulation (up to and including control of air quality), their 'visualisation' and their characterisation are at the heart of scientific, technological and industrial challenges. The actors act as if whoever had the right equipment had the knowledge and power for action. The question of right equipment is crucial (Vinck, 2006). The field of nanosciences and nanotechnologies would cease to exist without this material culture (Fogelburg and Glimell, 2003). Its progress and influence relative to actors seem to depend upon it.

The material culture associated with equipment in this field is characterised by the interweaving of instruments which combine manipulation and visualisation; microscopes are used as manufacturing tools. The field is also characterised by new combinations of instruments which go hand in hand with the importance attributed to interdisciplinarity and the articulation of fundamental research and technological development competencies. The closer the actors get to the nanometric scale, the more their equipment combines manufacturing, characterisation and theoretical comprehension. Simulation tools guide the design, selection of phenomena to be tested, the fabrication of objects as well as the control of objects and phenomena whose size prevents taking a direct measurement of the consequences of experiments carried out.

In addition to simulation tools, theoretical and conceptual tools also become important. Therefore, the researcher depends on both heavy infrastructure and 'immaterial' instruments: theoretical models, methodology and simulation, methods of optimisation, calculation tools. Whether they are fundamentalists or industrialists, they use these 'images' which are manufactured by intervention instruments on phenomena, the filtering of signals and their recombination. They produce them to access what they do and to make what they do according to others including colleagues, potential users of the results and those who commission the research and the general public visible to them. These images can be found in 'demos' and in communication formats (slide shows, articles and posters, works and presentation prospectuses) used by actors in this field.

laboratory creates phenomena and ensures stability for science through a constellation of instruments, interpretive procedures, phenomena and theoretical ideas that it defends; it ensures the closure of the knowledge network.

Local Instruments/Standardisation

Research tools are interesting to study when the technique is not yet established, when there is still a 'translucid box' (Jordan and Lynch, 1992) where local tests carried out by the actors involved can still be read.

Researchers often construct their own instruments themselves. For them, they are globally transparent and flexible. On the other hand, their dissemination

---

Box 7.10   *Technicians: the hidden part of the scientific iceberg*

Technicians often go unheard of, both in history and philosophy and in the sociology of science. This absence from the literature invites a question in itself.

The development of techniques is linked to social differentiation between scientists and technicians. Technicians and engineers who often develop instruments seem not to derive much profit from it; for example, they leave practically no trace in scientific literature.

Technicians and engineers, however, access other modes of fulfilment. They are invited to present their techniques in professional reviews or at technological forums. They develop instrumentation on the basis of new research programmes. Furthermore, observing laboratories reveals that technicians and engineers are also major authors; their literary production (report, technical documentation, diagrams, patents and arguments) are, however, often to do with grey literature and are a part of different professional dynamics.

---

is limited. The centrifuge, for example, 20 years after it was developed, was still only usable by its inventor, although it was considered to be an irreplaceable scientific instrument that many people have tried to copy. Researchers encounter enormous difficulties replicating instruments (Collins, 1974). In the reproduction of a laser by a team of researchers, publications describing the mechanism (internal reports and protocols) are not sufficient. It took numerous visits, demonstrations and informal exchanges between the inventor researchers and their adopters. Finally, they succeeded, but they never knew why the laser ended up producing the results they achieved. Its construction is dotted with milestones of discussion and adjustments; it is based on 'tacit know-how' whose transmission requires proximity. The knowledge transmitted is attached to individuals ('incorporated') who can carry out certain tasks without actually being able to explain their knowledge. The transfer of knowledge between laboratories is more like an apprenticeship than an exchange of information. In spite of visits and access to expertise on an ad hoc basis, the first trials to have produced the technique failed; the laboratory which had developed the technique did not itself understand all the parameters of the mechanism that it had developed. It took an extended period of learning, multiple contracts, trial and error as well as the construction of a relationship of trust between the researchers and their desire for the transfer to be effective.

Researchers also operate major investments (standardisation techniques and the standardisation of terminology, intercalibration, definition of specifications by the scientific community so that industrialists can produce instruments or products comparable from one laboratory to another) in order to make the comparison and aggregation of results from one laboratory to another possible (Vinck, 1992). When the industrialist enters the ring, his task is often limited to establishing a number of options and reproducing and distributing them. The centrifuge, for example, succeeded in being disseminated to different laboratories because a new version, adapted to new uses and possible improvements, was proposed by a manufacturer. Researchers got hold of it and continued to modify the instrument according to their needs. In certain fields, even very high technology fields, virtuosity of the technical bricolage or tampering on instrumentation is a trait of researchers' professional identity (Jouvenet, 2007).

Other instruments are produced by non-scientific activities. The galvanometer produced by researchers for industry, for example, was subsequently adapted to their activities at the request of scientists. The instrument is not scientific in itself but by the use that is made of it. Industrialists and scientists using the same instruments do not expect the same utilisation of them. Whereas the industrialist expected the galvanometer to produce exact results, the scientist was more interested in the sensitivity of the instrument. Researchers appreciate the mastery of their instruments because from that mastery, the facts themselves and the authority which results from it depend. Shapin and Schaffer (1985) thus analyse the practices and strategies developed by Boyle in order to create an instrument which allowed them to produce answers to sociopolitical and religious debates on the basis of his air pump. The laboratory created a new power, that of talking about facts with authority (Galison, 1987; Gooding et al., 1989). And yet, research instruments manufactured by industrialists tend to become black boxes for researchers which is both a source of advantages (not having to worry about the technical side) and disadvantages (loss of the control of phenomena and the records produced).[7] We also see researchers appropriating and using industrial instruments for other purposes in order to add them to some sort of research equipment and, therefore, construct a hybrid instrument (Hubert, 2007).

Once the tool is available (locally accessible and understood), it will often structure research activities: research problems and lines are reorientated in such a way as to make the best of the situation. Thus, in an endocrinological laboratory (Vinck, 1992) all researchers were invited to develop their research programmes around a new technique (the purification of B pancreas cells). The organisation of laboratories is linked to this technique which requires heavy investment: construction of a sterile laboratory and networks of collections of human pancreases; organisation of three-shift work so that a team is always available to deal with pancreases when they arrive.

Technical culture plays a role in the production and maintaining of scientific facts, in the permanence of research issues and scientific disciplines.

Conceptual Practices and Epistemic Cultures

Experimental practices have the merit of being able to be observed even when they are based on manipulatory activities. The question is raised, however, about the feasibility of ethnography of conceptual and theoretical practices, for example the mathematics.

Rosental (2008) observes that the logisticians employ considerable energy in demonstration activities, in particular in the preparation and execution of presentations at the board. This work explains the emergence of formalism (the practice of writing, discussions and transformations of logical expressions (Box

7.11) depending on critiques formulated). In the same way, conceptual work involves a broad material exercise in writing, correction, deletion and rewriting on paper, on the board and on computer. Some of the 'abstract' work can, therefore, be entered in its materiality (Latour, 1996). Rosental also reports on the emergence of a logic theorem by analysing stabilisation operations for debates, demonstrative practices and tacit know-how acquired during learning.

Box 7.11   *Relativistic and constructivist approach to logic*

Bloor (1976) looked at logic which he explained on the basis of the social institution. Concerning the works of William Hamilton, he proposes an interpretation of the development of algebra in terms of a conquest of independence by mathematicians in relation to religious institutions. Other authors insist on the role of tacit know-how and on the capacity of certain mathematical schools to impose their practices upon international networks. They analyse the coordination constructed around the theory of relativity, the influence of institutions in the orientation of research and the role of patrons and long negotiations depending on the strategic interests of the major manufacturers in the case of the arithmetic calculation as standard using floating decimal points (MacKenzie, 1993).

Concerned to report on the practicalities of the work, Pickering and Stephanides (1992) reconstitute the history of algebraic research on the basis of two pieces by Hamilton. They show that conceptual practice involves producing associations through which elements are connected to one another. The process of modalisation is made up of forced movements (reporting of the mathematician given the constraints imposed upon him/her by the notion and regulations already accepted) and free movements (which allows room for manoeuvre to the mathematician and to the social game).

On the basis of observations made in logic lessons and debates between logisticians on an electronic forum, Rosental (2008) reports on practices of inscription and of formalisation of logic expressions and their role in the dynamic of debates. The writings of logisticians do not produce the effects on readers that the author hopes for; the other logisticians do not replicate the form of reading desired by the author. What is obvious to some is not to others. Interactions between researchers are, therefore, necessary; they involve rewriting. Rosental thus shows the importance of the demonstration activities designed to 'make obvious' as well as the importance of associated tacit know-how in order to show, prove, demonstrate and trans-formalise.

The study of thought processes implies that collective dynamics are recomposed with their arsenal of data gathering (symbols, tables, graphs and notations), of instruments and of the practices which ensure the visual coherence of that data. These cognitive processes also involve the hands. They are more distributed than historical and philosophical reports would suggest. Abstractions are also the fruit of concrete, observable work. Indeed, scientists lay little store in their cognitive capacities (memorisation, rapprochement, distinction, succession) and often prefer to delegate this to objects whose action can be observed. Such is the case of the Munsell colour code with its holes used by pedologists, hesitant of affirming correspondences of colours as soon as they are separated by a tenth of a centimetre (Latour, 1995).

Most authors insist on the importance of tacit know-how associated with manipulatory practices of graphic and symbolic traces. This know-how is derived from apprenticeship and local cultures of research teams and specialised scientific communities. Knorr-Cetina (1999) reports on this and emphasises the diversity of 'epistemic cultures'.

The Theoretical Status of the Laboratory

Laboratory studies have contributed to a better understanding of scientific practices and the constructive nature of fact and knowledge. They also raised questions on the laboratory as both a social entity and a phenomenon. The question of the theoretical status of the laboratory has been raised.

Often seen as a site for the production of knowledge, which emerged in the nineteenth century, the laboratory has become one of the main focuses of attention of social studies on sciences (to the detriment of entities previously studied: scientific community, disciplines, scientific roles). This interest in the laboratory can be explained by the shift of attention of institutional aspects to the scientific work itself, its content and its processes. It also leads to moving away from an analysis which indexes phenomena according to social, cognitive, technical and natural factors, because these dimensions are associated in situ.

Notions of Laboratory and Experiment

Relations between experiment[8] and laboratory vary according to disciplines. Some laboratories are not linked to the realisation of any experiment (quality control laboratories or product certification laboratories in which trials and tests replace the role played by experiments) while certain experiments take place without laboratories (for example, simulations or experiments involving manipulating a representation of the world). Any laboratory that may be used is poorly equipped; it is only activated for the completion of the experiment and is coextensive to it. Certain simulations, however, require substantial amounts of equipment, such as the model of a port, or sophisticated technologies for manipulation of representations (big computers). The level of experiment, therefore, tends to become autonomous as in experimental psychology or with double-blind methodology in medical science. The experiment involves introducing a break between experimental subjects (that the experimenter strives to respect) and the interpretation of researchers.

In other situations, the laboratory covers the experiments (for example, in biology). This involves intervening on objects, transforming them according to a research programme, submitting them to a large number of tests and exploring particular effects. Instruments here have great importance. They produce

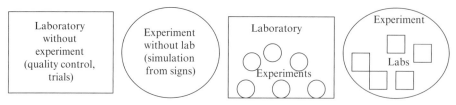

Figure 7.4   *Relationship between experiment and laboratory*

experimental effects. Here, there is no doctrine of non-intervention on the part of the researchers. Experimenters are identified with experiments; they transform themselves during the transformation process of objects covered in the study. Experiments in themselves have little importance; they are dissolved in the experimental process and are sometimes brought together for the needs of publication. The laboratory is a collective, social and political entity which is identified with the personality of the boss. It is the place for movement of material, equipment, data gathering and staff which is extended to include scientific cooperation networks.

Finally, the laboratory can be covered by the experiment. Such is the case in high-energy physics where the laboratory is simply one element among others for the experiment. The experiment is prepared over a period of years within an organisation grouping together several laboratories. At the end of the experiment, the data are divided up between the laboratories for analysis. Experiment orientates the work of these laboratories (Figure 7.4).

A Setting for the Consolidation of New Configurations

The laboratory is a theoretical notion; it is not only a physical space where experiments take place. It is a reconfiguration setting (Knorr-Cetina, 1981) or entities of nature and of society; an area which introduces a new phenomenal field of socialisation of elements of nature. It is an 'enriched' environment which takes objects from nature and installs them in a socialised environment. Thus, astronomy evolved from the direct observation of natural phenomena towards a mechanism for processing images. The laboratory is hereby reconfiguring social entities (researchers and teams) to make epistemic mechanisms out of them.

Laboratory studies show that the results constructed in laboratories become more consistent due to their local specification processes. The reconfiguration work within the laboratory derives benefit from resources and local opportunities in order to produce constructs that have more chance of resisting adversity once they are out of the laboratory. The laboratory is a protection setting for research projects in respect of the multiple contingencies which can affect them (Vinck, 1992). It is a niche which makes a specification process possible; it is an agent of scientific development.

Finally, the laboratory is a sociopolitical entity as much as an epistemic entity (Doing, 2004) where claims on knowledge are woven, along with the identity of the laboratory and control on work within and without. It is also a zone for the production of rules (Louvel, 2005) and anthropological demarcation (Hernandez, 2001).

Approaches and Criticisms of the Anthropology of Scientific Practices

Analysis of laboratory practices concerns several approaches which are summarised below, as well as the critiques that are made of them. Often these approaches combine and mutually influence one another.

- *The transversalist approach to scientific activity and socio-epistemological activity* (see conclusion of Chapter 6): laboratory studies here take intellectual and social structuring into account, along with the weight of cognitive and social factors. The authors here are keen to compare laboratories and reconstitute their own versions.
- *Ethnography*: this is not a conceptual approach but a methodological posture. It supposes the construction of a relation of strangeness in relation to local culture. The 'naive' observer who knows neither the language nor the custom attempts to approach and integrate the local situation under study, to understand what the people are doing and thinking, to get him/herself accepted, to have access to what is happening and to test it in concrete terms. His/her posture seeks to understand the shared and implicit cultural foundation and report on it in a language which is different from that of the locals.
- *Symbolic interactionism*: this is the study of reciprocal influence that partners of a situation exercise on respective actions (interaction is considered as a social situation in miniature). Field survey, in particular, has recourse to the observation of social processes and shows that scientific activity is made up of adjustments and negotiations. This interactionism focuses its attention on actors' perspectives (hence the description 'symbolic'), their interactions, the problem path and articulation work. The postulate is that scientific productions are the result of a social construction, produced by collective action and negotiations between actors in a given organisational context. The intellectual content is not differentiable from its organisational context (Star and Griesemer, 1989, Fujimura, 1987, Clarke and Fujimura, 1992).
- *Symbolic anthropology*: this concerns the study of cultural differences in a comparative perspective. Traweek (1988) thus compares communities of high-energy physicians in Japan and in the United States: leadership, organisation of work, design and construction practices linked to particle detectors. She demonstrates the contrast between American culture with its sporting analogies (whose pilot has learned to identify those competent and to develop a winning strategy) and the Japanese culture of the family (where each individual carries a responsibility for certain resources and where people's status is defined by their age). She also reports on the different affinities of researchers in relation to the manufacture of instruments: the Americans

prefer to shape their equipment themselves, unlike the Japanese who depend most particularly on industry. In a similar perspective, Knorr-Cetina (1981) introduces the concept of 'epistemic cultures' in order to shed light on the important differences from one discipline to another.

- *Constructivism*: the world of the human experience is constructed by categories of language and culture; it results from human work and social relationships. Scientific productions are local constructions which are explained by the processes that come out of them; they are situated and idiosyncratic. The idea of social construction was used to raise awareness and trigger critical examination. It is based on the hypothesis that X (the social construct) does not need to exist as it is today. It is not determined by the nature of things and is not inevitable. It results from a set of social forces and contingent elements. This idea contributed to questioning institutions which are presented as necessary staging posts. The question is about the object that we suggest is socially constructed (Hacking, 1999): is this an idea that we have of the object, of the categories and representations used to talk about it or is it the reality of the object? The fact that social institutions, technologies (Bijker et al., 1987) and innovation are socially constructed does not raise many objections. However, the statement that scientific facts or even objects in the natural world (for example, neutrinos), are constructs does provoke substantial controversy in scientific and intellectual worlds. Not all constructivisms, however, reduce their explanation simply to social processes alone. Constructs are explained in several ways according to the author: (i) as a result of negotiations and interactional accomplishment (Knorr, Lynch, Star, Fujimura, Clarke; (ii) as literary constructions and results of representational fiddling (Latour and Woolgar); (iii) as a local construction on the basis of means and resources which are available and depending on circumstances (Fujimura, Jordan and Lynch, Knorr-Cetina, Vinck); and (iv) as a result of local fact creation cultures (Knorr-Cetina, Traweek).
- *Ethnomethodology*: scientific facts and statements are considered as local emergences, practical achievements which are contingent and situated, which are inseparable from the course of the enquiry that produces them. They result from an interactional dynamic between members of the situation. The production of an agreement between people is explained by the procedures that they use in order to reach an agreement. The explanation is not to be sought on the social forces side which is hidden behind the actors, but in the present and local situation. The only pertinent elements for this explanation are those done and said by actors in the specific context in which they find themselves. Lynch (1985) identifies the scientific community by what can be observed locally: scientific research is nothing more than a tangible course for action and the conversations it observes. Observing work and conversations in situ allows it to take into account the indocility of materials and objects manipulated in the laboratory. It learns to master certain techniques used by researchers, note events which occur in the work, describe the sequential unfolding of activities, and record spontaneous

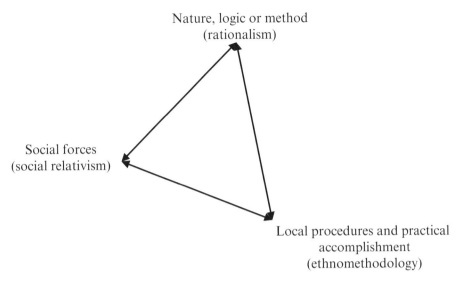

Nature, logic or method
(rationalism)

Social forces
(social relativism)

Local procedures and practical
accomplishment
(ethnomethodology)

Figure 7.5 *Principle of closure of controversies*

exchanges involving words and conversations. Its attention is drawn to gestures and words which emerge around issues which come up in practice, discussions about artefacts, expressions of researchers' uncertainty and conversations where work coordination is completed. Conversations are analysed as elements which make up the action. The ethnomethodological enquiry is distinguished from approaches which produce decontextualised reports (forms of reasoning, standards of behaviour, relationships of influence, forms of organisation).

Criticisms

Laboratory studies and constructivism have been subject to numerous and vigorous criticisms, the main ones of which are:

- **The local and contingent constitution of the facts**: that objects of nature are only made up of the process of enquiry and that the scientific facts and statements are inseparable from the courses of action which produce them is sometimes considered highly implausible. However, less radical versions can be considered to be more acceptable, in particular those which involve saying that entities discovered by science are given names, characteristics and techniques which are associated and are constructed. It is not stated that reality is a construct, but that it is locally specified and integrated into a sociotechnical space.

- **Local formation of scientific consensus**: most laboratory studies are limited to the study of a single laboratory or a small number of working situations. In reality, authors recognise that the formation of a scientific consensus often implies several laboratories or even the whole of a scientific community. Without losing the rigorous side of local studies of practices, it is important to be aware of processes whereby facts are stabilised by going beyond just laboratory sites and according to actors outside laboratories.

- **Ignorance of the context**: laboratory studies are limited to the study of what happens within the laboratory and ignores the institutional, political and societal context. Authors accept this fully and recommend the extension of this type of study to other sites of scientific production: scientific journal editorial committees, symposia, tribunals, regulation bodies (technical, bio-ethical and so on), science policy agencies, scientific cooperation networks, hospitals and industry. The idea is to monitor actors and objects (texts, instruments, materials). Laboratory studies also show how they are linked and interdependent in relation to outside actors; the laboratory is simply a link in the chain. It is important, therefore, not to overestimate the role of the laboratory; its role will vary according to the fields and situations studied.

- **Indifferentiation of the factors**: Gingras (2000) deplores the refusal to rank the order of factors which can explain the creation, circulation and dissemination of scientific facts and statements. This is problematic 'first of all because it is one thing to identify the heterogeneity of factors associated with the scientific activity, and it is another entirely to demonstrate their equal importance in the formation of a general consensus around the value of the products of this activity' (p. 193). This criticism is opposed to the constructivist position which consists of rejecting the suitability of an a priori distinction of social, cognitive, technical and natural factors when reporting on courses of action.

- **Forgetting the temporal dimension**: laboratory studies, even long ones, are limited to a given number of years. And yet, the history of laboratories is often long and its temporal dynamic is also interesting to analyse to understand the emergence of scientific productions. Also, scientific actors produce long-term objectives and regulations which frame their own action (Louvel, 2005). Their analysis should be carried out in the detail which characterises observations in situ.

- **Limits of the observer's naivety**: this strangeness represents a resource for the study of laboratories; in particular it avoids repeating the discourse of actors and allows us to ask questions on facts which seem natural to members of the situation. This strangeness and naivety also has its own limits, in particular in terms of access to understanding certain facts and gestures. The observer can, of course, ask for explanations but is not sure that s/he will obtain or understand the sense that the scientific actor attaches to it. The critique supposes that there would be a 'profound nature of researchers' activities' to which observers, without referring to the knowledge acquired in the domain studied, would not have access.

- **Absence of an explanatory framework**: the language of description proposed by Latour is made up of empty forms which add nothing to the words of the actors themselves. Its empiricism is thought to be positivist (Shinn and Ragouet, 2005) and textist (Bourdieu, 2004). These authors accuse Latour of mixing the ontological, epistemological and linguistic planes because of an abusive use of semiotics; Latour's intellectual project is indeed to cancel out such distinctions. Wanting to explain something by something else means distinguishing them a priori, which Latour refuses to do.

Conclusion: Extension of Laboratory Studies

To date there has been little ethnographic study of laboratories. In comparison to the very large number of laboratories which exist in the world, and their profound differences and transformations over time, the few studies of laboratories carried out allow us to identify only a small part of scientific activities.

Also, it has emphasised the extent to which it is important to study other sites of knowledge production in the same way: reviews, scientific committees, design offices, regulation bodies, technological platforms, associative movements (this work has been underway since the 1990s: Downey (1998), Keating and Cambrosio (2003), Vinck (2003)). It is about getting out of laboratories, multiplying the diversity of places observed and reporting on pertinent sociotechnical areas other than laboratories (for example, scientific cooperation networks).

Beyond the empirical studies inspired by laboratory studies, the constructivist approach has also seen substantial extension through the theoretical reformulation work carried out by Latour and Callon. Constructivism has led to the a priori indifferentiation of cognitive, natural, technical and social factors, it is important to propose a new form of terminology which avoids the traps of language which tend to limit researchers to categories of nature and society whereas, in reality, hybrids do exist. Callon and Latour propose a reformulation and extension of the anthropology of sciences in terms of actor-network and translation theory.

According to this model (outlined in Chapter 8), scientific statements are sociotechnical productions (material, formal and conceptual, inscribed, incorporated and materialised) which are dependent upon more or less wide and robust sociotechnical networks. Ideas cannot be reduced to social productions, since other texts and ideas are mobilised in their construction as much as instruments, entities of nature and the commitment of the body of researchers. They cannot be reduced to local production, since researchers strive to construct broad networks, so as to be able to support the universality of the statement produced. This model reports on the trajectories of objects, texts, people and the way in which they are associated in network format. The underlying methodological principles mean that the different elements and associations need to be processed in a symmetric manner, particularly taking into account humans and non-humans. Constructionist analyses of laboratories have led to a recognition of resistances that the material world offers in relation to the processes of social construction. Latour (1996) proposes incorporating them into the analysis of activities studied.

In the next chapter, we shall wind up our presentation by coming back to the issue of interfacing sciences, laboratories and society itself.

Notes

1 This situation is very different from high-energy physics, where the relationship to the research instrument conditions the involvement of a large number of researchers whose

specific and complementary competencies are required by the apparatus and by the phenomenon it studied. Work there is organised in a highly structured and centralised fashion.

2 See the compare presentation of 'instructions for fixing vascular profusions' and the ethnographical account on its implementation in Lynch (1985, pp. 69–74).

3 See Goody (1977, 1986) on the role of graphics, Eisenstein (1979) on the role of the printing works.

4 Concerning the scientist's job as an orator, for example, '*Noblesse oblige*', in Mulkay (1991, pp. 169–82) (parody of a Nobel Prize-giving ceremony).

5 Latour and Woolgar (1979 [1986]), Pickering (1984), Latour (1987), Traweek (1988), Vinck (1992, 1999, 2006), Knorr-Cetina (1999).

6 Galison (1987).

7 See analysis of the controversy between researchers on the preferential adoption of hot wire thermometry (flexible and understood by researchers) and laser thermometry (very precise but dependent upon industrial know-how) (Bagla-Gökalp, 1996).

8 The experiment is about isolating variables, exploring them and comparing them. It seeks to eliminate bias and subjectivity. There are few studies on experiments in the literature.

Recommended Reading

References appearing in other chapters: Eisenstein (1979) in Chapter 3; Knorr-Cetina (1999) in Chapter 6.

Bijker, W., Hughes, T. and Pinch, T. (1987), *The Social Construction of Technological Systems: New Directions in the Sociology and History of Technology*, Cambridge, MA and London: MIT Press.

Bowker, G. and Star, S. (1999), *Sorting Things Out: Classification and Its Consequences*, Cambridge, MA: MIT Press.

Clarke, A. and Fujimura, J. (eds) (1992), *The Right Tools for the Job: At Work in Twentieth-Century Life Sciences*, Princeton, NJ: Princeton University Press.

Downey, G. (1998), *The Machine in Me: An Anthropologist Sits among Computer Engineers*, New York: Routledge.

Galison, P. (1987), *How Experiments End*, Chicago, IL: University of Chicago Press.

Hernandez, V. (2001), *Laboratoire: mode d'emploi. Science, hiérarchies et pouvoirs*, Paris: L'Harmattan.

Knorr-Cetina, K. (1981), *The Manufacture of Knowledge: An Essay on the Constructivist and Contextual Nature of Science*, Oxford: Pergamon.

Latour, B. and Woolgar, S. (1979 [1986]), *Laboratory Life: The Social Construction of Scientific Facts*, Princeton, NJ: Princeton University Press.

Lemaine, G., Darmon, G. and El Nemer, S. (1982), *Noopolis. Les laboratoires de recherche fondamentale: de l'atelier à l'usine*, Paris: CNRS.

Louvel, S. (2005), 'La Construction locale des laboratoires. Approche ethnographique de dynamiques d'évolution de laboratoires académiques en France', PhD thesis in sociology, Université P. Mendès-France, Grenoble.

Lynch, M. (1985), *Art and Artifact in Laboratory Science: A Study of Shop Work and Shop Talk in a Research Laboratory*, London: Routledge & Kegan Paul.

Mondada, L. (2005), *Chercheurs en interaction. Comment émergent les savoirs*, Lausanne: Presses Polytechniques et Universitaires Fédérales.

Pickering, A. (ed.) (1992), *Science as Practice and Culture*, Chicago, IL: Chicago University Press.

Pontille, D. (2003), 'Authorship practices and institutional contexts in sociology: elements for a comparison of the United States and France', *Science, Technology and Human Values*, **28** (2), 217–43.

Rosental, C. (2008), *Weaving Self-Evidence: A Sociology of Logic*, Princeton, NJ: Princeton University Press.

Thill, G. (1973), *La Fête scientifique*, Paris: Aubier Montaigne, Cerf, Delachaux & Niesclé, Desclée De Brouwer.

Traweek, S. (1988), *Beamtimes and Lifetimes: The World of High Energy Physicists*, Cambridge, MA: Harvard University Press.

Vinck, D. (1992), *Du Laboratoire aux réseaux. Le travail scientifique en mutation*, Luxembourg: Office des Publications Officielles des Communautés Européennes.

Vinck, D. (2003), *Everyday Engineering: An Ethnography of Design and Innovation*, Cambridge, MA: MIT Press.

References

References appearing in other chapters: Shinn (1982), Vinck (1999) in Chapter 4; Latour and Fabbri (2000), Shinn (1988) in Chapter 5; Bloor (1976), Callon (1986), Feltz (1991), Galison (1997), Shinn and Ragouet (2005) in Chapter 6; Latour (1987) in Chapter 8.

Ashmore, M., Myers, G. and Potter, J. (1995), 'Discourse, rhetoric, reflexivity: seven days in the library', in S. Jasanoff, G. Markle, J. Peterson and T. Pinch (eds), *Handbook of Science and Technology Studies*, London: Sage, pp. 321–42.

Bagla-Gökalp, L. (1996), 'Le chercheur et son instrument', *Revue Française de Sociologie*, **37** (4), 536–66.

Berthelot, J.M. (2003), *Les Figures du texte scientifique*, Paris: PUF.

Bourdieu, P. (2004), *Science of Science and Reflexivity*, Chicago, IL: University of Chicago Press.

Cambrosio, A., Limoges, C. and Pronovost, D. (1990), 'Representing biotechnology: an ethnography of Quebec science policy', *Social Studies of Science*, **20**, 195–227.

Collins, H. (1974), 'The TEA set: tacit knowledge and scientific networks', *Science Studies*, **4**, 165–86.

Doing, P. (2004), '"Lab Hands" and the "Scarlet O". Epistemic politics and (scientific) labor', *Social Studies of Science*, **34** (3), 299–323.

Fogelberg, H. and Glimell, H. (2003), *Bringing Visibility to the Invisible: Towards a Social Understanding of Nanotechnology*, Göteborg: Göteborg University Press.

Fujimura, J. (1987), 'Constructing "Do-able" problems in cancer research: articulating alignment', *Social Studies of Science*, **17**, 257–93.

Garfinkel, H., Lynch, M. and Livingstone, E. (1981), 'The work of a discovering science construed with materials from the optically discovered pulsar', *Philosophy of the Social Sciences*, **11**, 131–58.

Gingras, Y. (2000), 'Pourquoi le programme fort est-il incompris?', *Cahiers Internationaux de Sociologie*, **109**, 235–55.

Gooding, D., Pinch, T. and Schaffer, S. (1989), *The Uses of Experiments: Studies in the Natural Sciences*, Cambridge: Cambridge University Press.

Goody, J. (1977), *The Domestication of the Savage Mind*, Cambridge: Cambridge University Press.

Goody, J. (1986), *The Logic of Writing and the Organisation of Society*, Cambridge: Cambridge University Press.

Hacking, I. (1983), *Representing and Intervening*, Cambridge: Cambridge University Press.

Hacking, I. (1999), *The Social Construction of What?*, Cambridge, MA: Harvard University Press.

Hubert, M. (2007), 'Hybridations instrumentales et identitaires dans la recherche sur les nanotechnologies. Le cas d'un laboratoire public au travers de ses collaborations académiques et industrielles', *Revue d'Anthropologie des Connaissances*, **1** (2), 243–66.

Jordan, K. and Lynch, M. (1992), 'The sociology of a genetic engineering technique: ritual and rationality in the performance of the "Plasmid Prep"', in A. Clarke and J. Fujimura (eds), *The Right Tools for the Job: At Work in Twentieth-Century Life Sciences*, Princeton, NJ: Princeton University Press, pp. 77–114.

Jouvenet, M. (2007), 'La culture du "bricolage" instrumental et l'organisation du travail scientifique enquête dans un centre de recherche en nanosciences', *Revue d'Anthropologie des Connaissances*, **1** (2), 189–219.

Keating, P. and Cambrosio, A. (2003), *Biomedical Platforms: Realigning the Normal and the Pathological in Late-twentieth-century Medicine*, Cambridge, MA: MIT Press.

Knorr-Cetina, K. (1995), 'Laboratory studies: the cultural approach to the study of science', in S. Jasanoff, G. Markle, J. Peterson and T. Pinch (eds), *Handbook of Science and Technology Studies*, London: Sage, pp. 140–66.

Knorr-Cetina, K. and Amann, K. (1990), 'Image dissection in natural scientific inquiry', *Science, Technology and Human Values*, **15**, 259–83.

Latour, B. (1979). 'Go and see: for an anthropological study of working scientists', *4S Newsletter*, 4 (winter), 18–20.

Latour, B. (1986), 'Visualisation and cognition: thinking with eyes and hands', in H. Kuklick (ed.), *Knowledge and Society Studies in the Sociology of Culture Past and Present*, Vol. 6, Greenwich, CT: JAI Press, pp. 1–40.

Latour, B. (1995), 'The "Pedofil" of Boa Vista: a photo-philosophical montage', *Common Knowledge*, **4** (1), 147–18.

Latour, B. (1996), 'On interobjectivity', *Mind, Culture, and Activity*, 3–4, 228–45 & 246–69.

Law, J. and Williams, R. (1982), 'Putting facts together: a study of scientific persuasion', *Social Studies of Science*, **4** (12), 535–57.

Lemaine, G., Clemeçon, M., Gomis, A., Polun, B. and Salvo, B. (1977), *Strategies et choix dans la recherche: à propos des travaux sur le sommeil*, The Hague: Mouton.

Licoppe, C. (1995), *La Formation de la pratique scientifique*, Paris: La Découverte.

Livingston, E. (1986), *The Ethnomethodological Foundations of Mathematics*, London: Routledge & Kegan Paul.

MacKenzie, D. (1993), 'Negotiating arithmetic, constructing proof: the sociology of mathematics and information technology', *Social Studies of Science*, **23**, 37–65.

Merz, M. (1999), 'Multiplex and unfolding: computer simulation in particle physics', *Science in Context*, **12** (2), 293–316.

Mondada, L. and Racine, J.-B. (1999), 'Ways of writing geography', in A. Buttimer, S.D. Brunn and U. Wardenga (eds), *Text and Image: Social Construction of Regional Knowledges*, Leipzig: Institut für Länderkunde, pp. 266–80.

Mulkay, M. (1991), *Sociology of Science: A Sociological Pilgrimage*, Milton Keynes: Open University Press.

Pasveer, B. (1989), 'Knoweldge of shadows: the introduction of X-ray images in medicine', *Sociology of Health and Illness*, **11** (4), 360–80.

Pickering, A. (1984), *Constructing Quarks: A Sociological History of Particle Physics*, Edinburgh: Edinburgh University Press.

Pickering, A. and Stephanides, A. (1992), 'Constructing quaternions: on the analysis of conceptual practice', in A. Pickering (ed.), *Science as Practice and Culture*, Chicago, IL: University of Chicago Press, pp. 139–67.

Ravetz, J. (1972), *Scientific Knowledge and its Social Problems*, Oxford: Clarendon.

Shapin, S. and Schaffer, S. (1985), *Leviathan and the Air-Pump: Hobbes, Boyle, and the Experimental Life*, Princeton, NJ: Princeton University Press.

Shinn, T. (1983), 'Construction théorique et démarche expérimentale: essai d'analyse sociale et épistémologique de la recherche', *Information sur les sciences sociales*, **22** (3), 511–54.

Star, S. and Griesemer, J. (1989), 'Institutionnal ecology, "translations" and boundary objects: amateurs and professionals on Berkeley's museum of vertebrate zoology', *Social Studies of Science*, **19**, 387–420.

Vinck, D. (2006), 'L'équipement du chercheur. Comme si la technique était déterminante', *Ethnographique.org*, **9**, February, available at: http://www.ethnographiques.org/documents/article/ArVinck.html.

Woolgar, S. (1980), 'Discovery: logic and sequence in a scientific text', in K. Knorr-Cetina, R. Krohn and R. Whitley (eds), *The Social Process of Scientific Investigation*, Vol. iv, *Sociology of the Sciences*, Dordrecht: Reidel, pp. 239–68.

Woolgar, S. (1988), *Science, The Very Idea*, London: Tavistock.

Zenzen, M. and Restivo, S. (1982), 'The mysterious morphology of immiscible liquids: a study of scientific practice', *Social Science Information*, **21** (3), 447–73.

8 The laboratory in society

· ·

Laboratory studies help us to understand the local construction of scientific output, but also its validation within the scientific community, its adoption by other researchers or socioeconomic actors, its translation into technological developments and its acceptance, dissemination and appropriation by society. Latour (1983) wrote 'Give me a laboratory and I will raise the world', but the mystery of the laboratory's power still remains intact. The laboratory reconfigures entities from the natural world and from society by inserting them into a sociotechnical assembly; it transforms human beings, creates new beings and produces visions of the world, discourses, instruments and collectives. However, simply creating such entities in the laboratory does not necessarily lead to the world being changed. It is therefore important to push beyond the laboratory and follow what happens to sociotechnical output outside of its walls.

Beyond the Laboratory

Beyond the laboratory, researchers depend on other researchers for the scientific validation of their local creations. If such creations are ignored or rejected, then they will forever remain local. If they are adopted by others, amended, integrated into new creations and, above all, if they leave a mark on posterity, they become universal truths, relating to nature and society.

Peer Review

Scientific output is subject to the critical appraisal of peers, who discuss it and suggest various modifications: a statement may be poorly backed up, data may be called into question, interpretations may be risky, or demonstrations not very convincing. Draft publications are submitted to colleagues and are passed around before they are published. Often, they are presented during conferences where they can be put to the test for the first time. Similarly, a PhD thesis is subject to several discussions and revisions before being submitted to an examining board. It is as if part of the review takes place in the corridors, during seminars and discussions behind closed doors or between colleagues before it is put before the official referees. Moreover, how is an examining board put together? Having already forged an idea of the value of the work and of the student, partly from informal discussions with colleagues and at seminars, PhD supervisors choose the members of the examining board accordingly. If supervisors feel that the work is fairly average, they will not bring in prestigious colleagues as this would undermine their own reputation. On the other hand, if the candidate is a brilliant student, they will use their influence to persuade higher-ranking colleagues to sit on the board: 'You'll see; it will be really worth your while'. In fact, the very composition of the board is the fruit of a first informal evaluation. The same applies before a draft publication arrives on an editor's desk[1] or before young researchers present themselves before a recruitment board. The discussions that lead to decisions about who is to sign a publication are part of the informal appraisal mechanisms (Pontille, 2003), of which still very little is known. Indeed, these mechanisms are sometimes even denounced when in fact they are an operative part of the review process.

The next step is the official evaluation: a viva voce, a competitive examination, the selection of proposals for a seminar and the examination of articles submitted to reviews. Here again, there have been very few analyses.[2] How are the referees chosen? What are the concrete procedures on which their evaluation is based? How are these evaluations negotiated within editorial teams and how are they integrated by the authors themselves?

Analysing the book reviews published in the *Revue Française de Sociologie* (French sociological review), Deloncle (2004) underlines the social factors of criticism. He shows that negative criticism is rare (less than 20 per cent) and is in line with a specific type of social logic: authors benefiting from a high amount of scientific capital do not attack other less well-endowed authors. Often those criticising and those being criticised are in an almost equivalent position. Negative criticism comes from commentators with less scientific capital. It works like a self-growth operator or like a confrontation between near equals.

Networks and Research Collectives

Scientific output is not necessarily subject to the aforementioned appraisal mechanisms. Many results, products and individuals are taken up by others (researchers, students, decision makers) within the framework of exchanges, cooperative work or contracts. This output circulates within spaces that contribute to the production of knowledge ('grey' literature, knowledge incorporated in individuals or made concrete through objects).

There are two opposing models of relations between research and the rest of the world. The 'confinement model' postulates an institutional separation between research and the rest of society (social demand, innovation). In this model, research is easily identifiable and visible. Researchers constitute a specific professional group, which is drawn into innovation processes according to needs. In the 'distributed research model', on the other hand, research activity is disseminated between actors for whom research is not the only activity. Rabeharisoa and Callon (2002) refer to the 'distributed research collective' to capture the joint dynamics of socioeconomic innovation and research. Analysing these dynamics involves following the movement of actors and multiple intermediary objects as well as reporting on scientific work in networks: mobilisation of worlds, creation of 'centres of calculation' (Latour, 1987), cooperative work and action on the world.

Inscriptions, intermediary and mobile objects

One part of scientific work consists in producing traces and inscriptions reported on through laboratory studies (design and adjustment of instruments, negotiation of interpretations, switch from one form to another). These inscriptions play an important role in the production of knowledge when they are (ibid.):

- **Mobile**: thanks to these inscriptions, phenomena can be moved around in time (to study them in the best conditions or when there is free time to do so) and in space (pulled out of the universe to be studied in the laboratory).

- **Immutable**: unlike ephemeral phenomena and samples that deteriorate, inscriptions are set and unchanging. It is possible to come back to them.

- **Flat**: it is very easy to spread them out on a desk and have a bird's eye view of them (unlike three-dimensional and opaque objects).

- **Proportional**: the scale of objects (whether galaxies or nano-objects) can be changed without altering the internal proportions; they can be held within several square decimetres.

- **Reproducible at low cost**: hence facilitating their mobility and dissemination.

- **Combinable**: thanks to the optical consistency of inscriptions, different aspects of a phenomenon can be joined and the phenomenon restructured.

- **Superimposable**: inscriptions of different origin and scale can be moved together, compared, superimposed and linked. Bringing them closer together can make structural effects or consistencies appear and lead to abstractions.

- **Insertable into texts**: texts and things can be brought closer together, compared and linked to produce semiotic homogeneity.

- **Ready for geometrical processing**: projected onto paper, the objects under study (whether small or large, from the natural world or society), can be measured and incorporated into the world of mathematics. This makes them easier to handle than when they are made up of words or three-dimensional.

Scientific work also focuses on intermediary objects (Box 8.1) extracted from various worlds and formatted for use by the laboratory (fossils, photos, field notes, samples, data and so on) or created in the laboratory (models, probes, cross-sections, instruments, transgenic rats and so on). Researchers strive to turn these into immutable and combinable mobiles using preservation and standardisation techniques (herbarium formats and histological sections) negotiated between researchers, collectors of samples or data (field naturalists, general practitioners supplying clinical cases, investigators and so on), amateurs or professionals. Some intermediary objects are more mobile, immutable and combinable than others, notably when they are digitised and can travel around computer networks. The design, production, dissemination, preservation and use of these objects occupy researchers and take up a lot of the resources in scientific cooperation networks (Vinck, 1992).

Box 8.1   *Intermediary objects in scientific cooperation networks*

Documents: lists of data, articles, reports, research proposals, questionnaires, photos, patents, protocols, order forms, tapes and so on. These enable researchers to be in contact with other scientists and all those interested in their work (clients, teachers, industrialists and so on). They represent their authors and the natural objects themselves; they are authorised to express what these entities do while their authors set themselves up as the legitimate spokespersons.

Products, reagents, materials, specimens and samples (including animals or human body substitutes called phantoms). The accessibility of research material affects the social organisation of research, the development of a given field and its cognitive aims (see Oudshoorn, 1994 in the case of urine and sexual hormones).

Instruments: as part of the laboratory infrastructure, reflecting incorporated know-how and acting as the spokespersons for conceptual and theoretical approaches, instruments are associated with more or less restricting specifications for use. They are accompanied by texts (instructions for use), other objects (reagents, utilities) and people (demonstrators, repairers and experienced users).

The dynamics behind the exchanges and scientific cooperation also involves the movement of people (for example, researchers and technicians), and hence their incorporated scientific, technical and organisational skills. Furthermore, scientific networks also involve external partners, supplying data and equipment.

Mobilisation of worlds and centres of calculation

The laboratory is not a local entity that is closed in on itself, where theory meets nature, and cognitive and social factors join forces. It is connected to networks that promote and help to transform it. The identity and power of the laboratory are also defined by these networks. Science studies thus take us beyond the laboratory; they reposition the laboratory as a centre of accumulation and transformation of all kinds of elements.

According to Latour (1987), the laboratory draws its strength from its ability to mobilise worlds, in other words its ability to attract objects from various universes (museums, data or sample banks, files, centres of calculation and so on). This mobilisation involves the construction of networks enabling the laboratory to draw these objects towards it: organisation of expeditions, launch of interplanetary probes or organisation of data collector networks. It brings together entities from far-reaching horizons and creates the possibility of making a switch from local, indigenous knowledge to universal knowledge. From this viewpoint, the print shop was an agent of scientific change (Eisenstein, 1979). It made it

possible to bring together old work and field surveys, and to compare and summarise information. It facilitated the dissemination of common references and
helped to make science cumulative and universal (within the limits of the texts'
dissemination).

From this viewpoint, the laboratory is a micro cosmos that mobilises a
macro cosmos via relatively long, robust and extensive networks. The power of
the sciences and the robustness of scientific statements are linked to worlds that
are mobilised and transformed in the laboratory, and which support the texts produced there. Inscriptions, sometimes representing vast universes, can be studied
by researchers in the laboratory and hence dominated either visually or using
local instruments. The hold that researchers have on the universe stems notably
from this ability to bring together optically consistent traces, representing separate events in time and in space. Work on inscriptions (comparison, combination,
classification and so on) involves keeping together large quantities of inscriptions
in order to produce knowledge about an entire population (fauna, atomic table,
companies, households and so on) (Desrosières, 2002). New combinations are
also invented and tested in the laboratory, and used by researchers to form new
entities. Thus, by counting the traces in bubble chambers, physicists can isolate
the characteristics of an elementary particle; based on the data from a survey on
a given population, sociologists can identify groups and social representations.

Research collectives

The availability of world mobilisation networks does not always ensure the dissemination of knowledge and other laboratory products. The newer the knowledge and products are, the greater the risk that others misunderstand them and
find them difficult to adopt. They can only be understood by those having made
them or by those sharing similar skills and resources. They correspond to 'short
networks' of local and incorporated knowledge as opposed to 'long networks' of
knowledge that is codified and mobilised at a low cost.

Their dissemination also depends on the construction of networks where
the skills, instruments and language used are similar. Sometimes, the skills are
widespread (thanks to the harmonisation of teaching content), the instruments
comparable (thanks to technical standardisation and the commercial strategies
of manufacturers) and the samples equivalent (thanks to research protocols
defined jointly within international scientific societies and to their publication).
This equivalence between laboratories is the result of work to intercalibrate
instruments, control quality (via knowledgeable societies in clinical biology, for
example), define new methods and validate them, train young colleagues and integrate them into other laboratories, as well harmonise practices and create reviews
so that new approaches can be disseminated.

Distributed research collectives also spring up around laboratories. They
are composed of researchers, knowledgeable users and concerned non-specialists
who interact and pool their knowledge. Their skills become hybridised while
together they define the problems and build shared knowledge. These collectives are hybrids in so far as their members may not share any common values

or visions or may not feel a sense of solidarity and belonging to a specific group. Referring to objects and technologies is enough to bring them together and create interdependence.

According to Knorr-Cetina (1999), the production of knowledge is indissociable from these distributed research collectives, which influence protocols and the flow of materials and texts. In the case of high-energy physics, hundreds of dispersed researchers and engineers (from different countries, disciplines, private and public organisations) interact around an instrument and an experiment over a period of several dozen years. Moreover, an individual researcher often only participates in a fragment of an experiment and is just one involved agent among many others. The individual researcher cooperates and co-signs articles (sometimes articles that already have several dozen signatures). These distributed research collectives and the way they work vary (centralisation/decentralisation, flexibility/bureaucratic formalisation, homogeneity/multipolarity).

These epistemic communities are organised to produce codifiable knowledge. They are different from communities of practice (Wenger, 1998) made up of individuals engaged in similar practical activities with an interest in sharing and developing knowledge associated with these practices together. Whether epistemic or practice based, a collective (Rabeharisoa and Callon, 2002) can become a community (Amin and Cohendet, 2004), if a sense of specific identity develops. Scientific cooperation networks correspond to these different situations, ranging from the building of a shared vision of a problem to the harmonisation of research practices.

The action of research on the world

The laboratory is a micro cosmos where social and natural worlds are redefined and new beings created (ideas, objects and so on) and disseminated. These new beings have an influence on awareness, knowledge and the world of business, health, public policy, students and the media. In the laboratory, researchers, for example, develop beings that they present to colleagues and industrialists. This was the case when a group of biologists developed transgenic rats with high blood pressure. They spread the news about the rats in their publications and made them available to other laboratories. Several years later, without any concerted decision being taken, many laboratories were looking at the problem of high blood pressure in a different light. In fact, they had all adopted the new working 'tool', that is, the transgenic rat, and had redirected their research strategies accordingly.

The models and texts produced in the laboratory, as well as the people who are trained by researchers (future citizens, industrialists, consumers, teachers, managers, politicians and so on), help to redefine political agendas, as was the case with global climate change. The laboratories looking into global change issues trained experts who then became assistants to policy makers and set up new hybrid institutions such as the International Panel on Climate Change (IPCC). This body defines both scientific and political agendas (see also the case of health economists in Great Britain: Ashmore et al., 1989).

Scientists also build instruments that they show to colleagues and industrialists.

These instruments channel scientists' know-how and act as spokespersons of their concerns and working methods. They extend the laboratory's action in society, as in the case of laboratories that coordinate their work so as to head industrial developments in a certain direction. In the case of automatic systems for the classification of electrocardiograms, researchers were even able to impose new technical standards on industrialists, via standardisation organisations. In the case of climate change, scientists have produced global warming assessment models, which have become politically hegemonic. Conversely, the design of these models also reflects the political aspirations of researchers, that is, the setting up of forecasts in order to sway public policy makers.

Finally, money comes out of the laboratory in order to buy the products it needs. Researchers can actually direct industrial developments and strategies via the orders they place. In the case of 'Boron Neutron Capture Therapy', laboratories grouped together and were thus able to persuade industrialists to produce a product that they needed for research. By agreeing, by defining the specifications of the reagents needed for their research, and by developing a new therapeutic method, they created the bases for a new market that they pushed industrialists towards.

Researchers, therefore, do not just work on the design of new knowledge, instruments, products and expertise; they also contribute by creating the demand for and dissemination of such things. They do this by extending their networks downstream of their activity. Sometimes, researchers act directly on the world (for example, through hospitals or companies) to make it possible for the methods developed in the laboratory to be applied there. The movement of students on company placements or carrying out professional theses constitutes one of the vectors of transfer of knowledge from laboratories to the world. The development of measuring instruments, metrological methods and new standards (for example, concerning the potential health risks of disseminating nanoparticles in the environment) helps to transform the world into an extension of the laboratory so that the methods developed within the laboratory can be applied outside of its walls.

Knowledge is also co-constructed by actors, in which case academic research is only one link in the chain among others. It is swept up in a movement of which it is no longer the driving force. This is sometimes the case of research collectives such as orphan and affected groups (Rabeharisoa and Callon, 2002).

The open source 'community' is another example of a research collective that has emerged outside of existing sociotechnical frameworks ('orphan groups'). Within this community, links have been woven between academic research, the industrial world and users of computer products as the collective has attempted to redirect innovation paths. Its aim is to solve some, at times, sophisticated technical problems, find quality solutions and share these. To begin with, this community was a collective of individuals faced with the same technical problems and the monopolising practices of a major corporation (Microsoft). Exchanges developed until a coordination body was set up to assess, classify, preserve and disseminate the contributions of each member, providing them with recognition through symbolic rewards. This body defined the rules and set an objective to develop alternative software. The knowledge produced was formalised and codified; the collective changed from being based on practices to being epistemic. It unlocked the innovation framework imposed by Microsoft, helped new firms to enter the game and launched new academic research. Its members were dispersed across the world and in various organisations (industry, public research, end users, consumer associations and so on). Labour market sociology in this field also shows that the professional paths of the individuals in such communities evolve: from being seen as amateurs they gradually come to be viewed as recognised professionals. Activities may be on a volunteering or trade basis, or they may be academic, industrial and militant. The distributed collective is a suitable entity for innovation. Californian bioinformaticians and their local associations, or other associations bringing together professionals, researchers and non-specialists (around a disease), have set up alternative research directions, along with their own research systems and ways to discuss results. They are examples of orphan groups (Rabeharisoa and Callon, 2002 for myopathy; Epstein, 1995 for AIDS).

Other research collectives ('affected groups') have emerged from this awakening to the excesses and unexpected effects of sociotechnical innovations: for example, the potential effects of relay antennas on the health of those living close by. These have propelled knowledge production dynamics and encouraged researchers to redirect R&D. This is what happened when people living close to a mine observed an abnormal number of cases of child leukaemia and alerted the specialists (who were not interested to begin with). Surveys were carried out and experts called in. The collective group grew and become organised. It started several court proceedings. After several years, a new syndrome was recognised and included in the official classifications.

Actor-network Theory (ANT)

We have just seen how knowledge production and innovation arise from the construction of sociotechnical networks and, therefore, from heterogeneous engineering (Law, 1987). Reporting on the robustness of this production, Callon and Latour developed an epistemological and methodological conceptual framework (Chapter 6) and an anthropology of laboratories (Chapter 7). This conceptual framework is based on the principles of symmetry and association to explain the differences produced by science and technology.

In accordance with these principles, they put together a repertoire enabling them to talk about technical and social aspects using the same terms. Callon (1986) proposed this translation repertoire because terms like truth, nature, rationality and so on, are used by actors. 'Translation' is a general process according to which a social and natural world is shaped and stabilised. It includes several dimensions: problematisation, 'interessement', enrolment and mobilisation of allies. It is the basic mechanism behind the setting up of relations that form networks. This new concept was a fundamental notion for sociology (Latour, 2005).

Problematisation

When engineers launched projects to develop an electric vehicle in the 1970s, they first produced discourses in which they traced the borders between what is a problem and what is not, between technical questions and economic issues, between what is known (that is, what can be applied) and what is not (that is, what requires research). They outlined a set of realities, seen to be irrefutable, and situated the problems to be solved. They built a new reality through words and arguments while deconstructing another. They defined and set out entities and new relations and a new conception of reality (the current problems, the definition of a new world, and the definition of a new sociotechnical world to come); they proposed a 'problematisation'. They attempted to impose this definition of reality using arguments and a text that would serve as an authority.

Problematisation occurs in the first stages of research, during identification of the objects to be studied, the issues to be solved, the logical links to be set up and the appropriate approaches to be adopted. The actors link up the contents and skills to be brought together and identify the groups (disciplines and companies) to be mobilised to work on these problems (sociological links). They simultaneously define the contents and contexts. These links are 'sociological'. Each actor has his/her own problematisation, which is linked up with other actors or other problematisations.

The relation set up between these problems is called 'translation' because it operates between distinct registers that it brings together or renders equivalent. Fields of heterogeneous activities are brought into contact with one another and a passageway is proposed between them. For example, researchers working on fuel cells translate 'energy crisis' by 'electric car', which itself is translated by 'fuel cell' and 'electrochemistry'. Heterogeneous elements are thus linked: political programmes are linked to theoretical debates, laboratory entities (electrodes) are combined with macro entities (France and the evolution of society, climate change and so on). Such associations between heterogeneous elements can be observed during scientific controversies as well as in research and innovation projects. They stem from scientists just as much as from legislators, industrialists or consumers. They help to redefine society and technology, as well as science and nature. Legislators define rules to protect inventions; industrialists concoct strategies involving public laboratories; associations of patients suffering from an illness set up media-related actions to reach out to scientific institutions.

Nevertheless, the translation operated in a given problematisation is only based on conjecture. It indicates the relations and displacements (of problems, actors, terms, world vision or resources) to be implemented. It is the construction of a hypothetical reality. However, it does have its own consistency, that is, that of the discourse attached to an actor. If this hypothetical construction is taken on board by others and integrated into their own constructions, it becomes more real, consistent and robust. Its solidity can be explained by its 'selective integration' (Knorr-Cetina, 1981) into a new production (scientific, technological or other). It is negotiated; some are opposed to it while others remain indifferent. It is thus a question of following and reporting on the negotiations and confrontations as well as on the resulting displacements. The first divisions are modified and the structure of relations transformed. Some problematisations are consolidated in the process of being adopted; others fall apart and disappear.

Problematisation proposes a relation and a displacement. It formulates problems and demonstrates that, in order to achieve their objectives, the entities identified must operate this displacement. For example, through the title and text of a publication, a biochemist may build a problematisation in which a problem (cancer – the problem of a doctor, who would like to fight against it but does not have the appropriate tools) is linked to a new method (developed by the author of the text), which the reader/doctor is invited to switch to if they want to achieve their objectives. Such is the fundamental translation mechanism: it proposes a relation between activities, interests and problems.

Interdefinition of entities and obligatory points of passage
By setting up a series of links between problems, problematisation defines obligatory points of passage (an innovation or an innovator). It shows the detours to be granted (for example, the idea of the automotive industry producing electric vehicles (Box 8.2)) and the alliances to be sealed for example, with the help of electricity producers). The entities (either human or non-human),[3] are defined as imprisoned in their existence by obstacles (for example the combustion engine is bound to be phased out). Problematisation defines the identity of an alliance system and the displacements to be operated in order to get around the problems coming between these entities and what they want.

Problematisation is also an attempt to redefine associated entities and their properties (for example, the ideas of cars being electric from now on). These are not necessarily givens; there is no a priori defined list of either entities or their properties. Problematisation is an operation that consists in (re)defining these. An entity's identity, properties, stability and so on, are redefined during the translation process via the relations built between entities. Problematisation is an interdefinition; the entities define each other mutually. Problematising therefore

---

Box 8.2   *The electric car (1)*

Callon (1986) shows how, in the 1970s, the EDF French electricity authority problematised and shaped the evolution of industrial society for its own profit. EDF redefined the social world and its evolution (the end of the consumer society, the search for quality of life, the end of the petrol-driven car, symbol of this past society, and the inevitable arrival of the electric car). The company also defined the state of technologies, the corresponding sectors (a black box composed of processes, laboratories and the qualified and concerned industrialists), and the chronology of evolutions to come. It defined the products that were bound to be produced by industrialists and wanted by consumers. It defined itself as an obligatory point of passage. EDF also redefined the role of Renault: a company whose future would consist in manufacturing the chassis of electric vehicles. Was this a dream or reality? It is impossible to say since this is precisely what the actors *were* fighting about.

consists in hypothetically establishing the identity of an entity and what binds it. The result is an 'actor-world': a set of problems and entities within which one entity renders itself indispensable to the others by building obligatory points of passage.

Interessement and enrolment

The second dimension of the translation process is 'interessement', a process according to which the identity of other entities is imposed and stabilised, notably alongside the redefinition of their interest (which is not normally a given, any more than the other properties). It is a question of forming a network of alliances defined by the problematisation.

'Reality' is nevertheless a process that passes through successive states of making or non-making depending on the events and tests undergone by the established associations. If a specific type of problematisation is taken up by other actors, it becomes stronger and more consistent. If a publication is read and cited, it becomes more real. If a research project put forward to the science council is approved, albeit with a few modifications, and included in the science policy programme, its reality is enhanced. The same applies if it is translated by a subsidy agreement and the recruitment of a new researcher.

Actors propose problematisations, but also strive to make them real, by acting in such a way as to make others want to take them up. They set up interessement devices in order to detach entities from their former attachments and get them to enter into the relations projected in the problematisation. It is a question of displacing entities: so that laboratory X focuses on a given problem in collaboration with Y, so that virus Z can be attenuated and the Science Council agrees to a subsidy. Scientific argumentation is an interessement device (Callon et al., 1986; Callon and Law, 1982), but there are many more such as diverting an object or an animal from its normal path in order to get it to go through a pre-defined obligatory point of passage (OPP).

It is also a question of breaking the links of the entity to be displaced in order to set it within new associations. The nets used to capture animals, the moral discourse used to bring auditors back on the right track, a whole range of sensors, and money, are all interessement devices. Similarly, some technical devices (for example, e-mail) or types of organisation (for example, Club Med or certain companies) help to break up the social links of human beings and remodel their identity by holding them in new webs of relations. These interessement devices should therefore be identified and their action reported on. They help to understand what holds together these new structures (logical, sociological and so on) or the translations proposed by the actors.

Beyond interessement, there is 'enrolment'. This is a mechanism according to which a role is defined and attributed to an actor, who then accepts it. Enrolment is interessement that has worked. It makes it possible to report on and understand the establishment, attribution and transformation of roles. Unlike functionalist or culturalist sociologies, where society is made up of a repository and a combination of roles and role holders, the sociology of translation neither involves nor excludes any pre-established role. The role is constructed at the same time as the actors are enrolled.

Chains of equivalence and mobilisation of allies

Mobilising allies consists in making non-mobile entities mobile. By designating spokespersons and by linking up a cascade of intermediations and equivalences, many entities are replaced by a spokesperson, chosen by those in whose name the spokesperson is to speak (a delegate) or created by the researcher (a representative sample). Following these transformations, a multitude of entities is replaced by a handful of spokespersons able to displace the whole set.

Researchers get nature to take a detour via their laboratory using a series of translation operations. They mobilise selected and questioned spokespersons (samples, representations and so on). Their statement is recorded, compiled and compared in the laboratory. By selecting these spokespersons, researchers reduce the number of interlocutors with whom they have to interact. If they can set up a faithful spokesperson instead and in place of the multitude, the situation becomes much easier to control. The multitude is displaced, simplified and punctualised (that is, transformed into a point). The notion of 'translation chain' describes the series of displacements and the setting up of equivalences needed to produce a statement or an object. Hence, a scientific publication summarises and displaces a series of texts that it cites, along with objects manipulated in a laboratory, but also human beings (researchers, technicians, competitors, financial backers and so on). The statements in the text translate and refer to other statements, objects or actors that they summarise and link up.

Via this translation process, actors produce asymmetries and structure a network that they attempt to hold together. They enrol elements in order to consolidate the small provisional asymmetries inscribed in texts or material devices, incorporated within individuals or set out by a new institution. They create irreversibility and stabilisation. The entanglement of translations outlines a sociotechnical path that reduces the amount of room for manoeuvre enjoyed by the entities involved. When a translation is successful, it takes the shape of a restrictive network for these entities. However, the entities mobilised can always escape: the elements on which a given reasoning is based can fall apart, social habits can change and machines break down.

Actor-network

Heterogeneous combinations arise from scientific and technological work in the form of statements, technical devices, incorporated knowledge organisations or new worlds. These combinations are linked to sociotechnical networks which, when they act as a new actor, are referred to by Callon (1986) as 'actor-networks' (Box 8.3).

In some cases, the translation postulates new entities (stemming from nature or society) and attempts to bring them into existence. Their list is constantly subject to change: some emerge while others are redefined like Franklin Delano Roosevelt, who, from a politician ignorant of what physicists were up to, was redefined by Einstein as the 'President who wants the atomic bomb'.

Box 8.3   *Actor-network theory and causal explanations*

A cause acts as a force: the closer one is to it, the more active it is. Thus, the more scientists apply the right method or the more compliant they are with the ethos of science, the more likely they are to produce valid results. Put differently, the closer scientists are to a social group, the more their results are influenced by this group.

Supposing there are explanatory causes is like supposing there is an underlying space of forces (social or natural) that explains appearances. In sociology, this approach leads to processing survey data so as to underline the explanatory factors that best capture the data. The set of causes forms the explanatory structure, which takes the form of a space whose number of dimensions corresponds to the number of explanatory causes selected (factorial correspondence analysis and main component analysis methods). Data classification techniques thus consist in building classes of objects based on distances (proximities or similarities). Such analyses reduce the relations between entities to several selected explanatory dimensions.

The **actor-network theory** is radically different from this approach. Here, distance is linked to the path followed. For an underground user, the distance between two places in the city is linked to the number of stations separating them rather than to their Euclidean distance. If there is a direct connection, the distance is a short one; if there is no connection between the two points, the distance is longer. The actor-network theory describes associations and follows the sequences of translation. It does not suppose an underlying space for these sequences. It does not suppose a relation between entities who are not linked by an identifiable path. The path is the only thing that counts.

The points (objects, words, texts, individuals, groups and so on) are actors if they are associated with other points. The more a point is associated with other points, the more it is considered a potential actor. Interactions define this point as an actor and an attractor. Its importance stems from the weight of its relations, whose specific and irrefutable character must be respected.

The universes linked up by an actor can be heterogeneous. In other words, each one can have a different analysis or reference grid. To capture the relations between them, the survey must move around. It starts with a hypothesis (problematisation) stemming from the associations established by point A in order to define B, then checks whether the associations of B confer the role that it attributes to itself on A. Using B's associations to define A, the survey comes back to A and so on and so forth. Thus proceeding in successive iterations, it does the same for each relation. (Note: the points can be individuals whose discourse refers to each other, words associated with other words in texts, technical devices linked to people, texts and other objects.)

The application of these principles in order to analyse bibliographic databases and patents as well as to analyse texts is explained in Callon et al. (1986).

Actants are themselves networks. Studying an actant (a neutrino, a scientific law, a diagnosis kit, a laboratory and so on) involves following its construction. Its meaning comes from the associations created and its identity depends on the translation operations. Because networks change, so too do identities. There is no immutable actor (whether one is referring to a pressure group, a social class, an individual or an elementary particle). The identity of entities depends on the structural weight of the network. The meaning of a statement, its force and its ability to convince, for example, depend on the chain of translations and the reference created by the network. The power of conviction, just like the efficiency or robustness of a technology, the legitimacy of an argument or the social accept-ability of a new technology, depend on the morphology of the networks and the robustness of the translations of which it is composed.

Extending networks involves mobilising many different entities and building relations between them to form a new actor-network. Punctualising the network does not mean that there is internal homogenisation. Some elements make it pos-sible to maintain this diversity ('boundary objects': Star and Griesemer, 1989) and 'mediators'.

Although translations and networks can be robust, they are nevertheless trial runs that may or may not work. Sometimes, they fall apart: the spokespersons are denounced; the actors turn back to their initial liaisons; the instruments fall to pieces; the theories prove to be inconsistent. Networks and spokespersons can be called into question. Entities can resist the definition imposed on them and act differently. New translations can divert them from the obligatory points of passage imposed on them. Liaisons can fall apart and networks become dislo-cated and non-realisable. The result is that the description of reality begins to fluctuate (Box 8.4), as is the case with some innovative projects: the right techni-cian might leave for a better paid job, a bolt might give way, a customer change their strategy and social movements denounce the excesses of innovation. In the Aramis underground story (Latour, 1996), descriptions from one actor to another stopped being superimposed on each other when the network became less real. Controversies broke out and the representativeness of the spokespersons was questioned, discussed or scorned.

Furthermore, the construction of these networks can be limited by other networks, objections, rules or technical devices that restrict the scope of accept-able translations (for example, the mechanisms for appointing spokespersons or setting up a representative sample), the spaces of circulation (statements, instru-ments and skills) or the distribution of rights (property rights, rules relating to confidentiality and so on).

Reporting on asymmetries

Refusing to take as a starting point the distinctions between content and context, or science and society, does not mean that everything amounts to the same thing. On the contrary, actors strive to differentiate between things (between truth and error, knowledge and belief, human and non-human and so on). The construc-tion of these differences occupies actors. Researchers, for example, do their best

---

Box 8.4   *The electric car (2)*

For several years, Renault was subjected to the problematisation set up by EDF. They got through it. Everybody recognised that the individual car was bound to disappear and that the heat engine was polluting and costly. The question was how to repudiate the actor-world built by EDF? How could Renault deconstruct EDF's problematics and open its black boxes (that is, the electrochemical knowledge of this electricity-monopolising company)? Renault resisted and strove to dissociate the elements associated by EDF. This led to investigations, testing of links, a search for new allies and reproblematisation. For Renault, it was about transforming the reality imposed by EDF into fiction.

EDF associated the increase in petrol prices with the decrease in car demand; Renault showed that the facts contradicted this association. Everything was going up: the price of petrol, the number of cars purchased, the fight against pollution, and the overpopulation of cities. Renault retranslated social demand: consumers wanted an individual car, speed, comfort and good car exchanges. These were things that the electric car could not offer, and so there was no market for it. Renault pulled apart the definition of society built by EDF and replaced it with another. Similarly, it pulled apart the technology built by EDF: the car maker questioned scientists and engineers, and re-examined the state of electrochemistry. Renault then found that it could develop its own engine using electronics and that it was impossible to make storage and fuel cells for electric cars. A scientific controversy emerged. The network built by EDF became fiction, a dream jotted down on paper. Renault redefined problems and relations, and gained the interest and enrolment of new allies (consumers and electronics).

---

to distinguish between valid statements and personal beliefs. Engineers develop machines that stand out from those of their competitors. Philosophers strive to make an argument solid and intrinsically consistent. These asymmetries that they build are all the more solid because they are based on robust and extensive networks. If a geographer's statement carries more weight than that of the village elder, this is not because the former has more intelligence and method, it is because his/her statements do not rely on the same networks. The latter knows the region while the geographer bases his/her hypothesis on a stack of multiple traces. The outline of an island drawn by a local on the sand is worth nothing to the local as he/she knows the island inside out. For an explorer, on the other hand, this sketch is worth everything. It is an intermediary object that makes all the difference, especially when copied onto paper (which, unlike sand, is a mobile and stable medium). The faithfulness and stability of the drawing on paper is also important when it comes to preparing new trips. The drawing is passed around and enrolled in a network (commercial); it makes it possible to prepare new routes and explore different scenarios. It is rendered homogeneous by being drawn according to the correct longitude and latitude. Just by looking at the drawing, new things can be learnt about the island by taking measurements and comparing it to other drawings of islands. It is no longer necessary to be on the spot to find out more about such places. The fact established by geographers becomes universal, not because it is rational but because it is reproduced and adopted by others who use the same codes and instruments, unlike local knowledge that is not disseminated, compared or connected with other knowledge. Thus, for Latour, there is no such thing as a 'Great Divide' that turns Western scholars into superior beings, there is simply a piling up of multiple differences (Latour, 1988, 1993):

> To convince a fellow physicist, it may be necessary to invest several million dollars and years of work. Who can afford such strength of conviction? . . . A 'theory' apparently only requires a paper and pen. But the craftsmen able to produce science at such a cheap price were supplanted a long time ago. To issue the slightest credible opinion in particle physics or climatology, you have to have powerful computers and huge data bases. (Latour, 1982, p. 41)

Science, Technology, Innovation and Society

Having moved out of the laboratory to explore distributed research collectives and sociotechnical networks (actor-networks) that reflect the robustness of scientific output, we shall now turn our attention to innovation. Sciences, technologies and societies are often thought of in linear terms: science makes the discovery, technology applies it and society follows. Social studies of science and technology show that things are much more complex, that the processes follow a zigzagging path and that they lead to co-constructions.

Science and Technology

Observers agree that there is no simple relation between science and technology. Technological development cannot be reduced to the application of scientific discoveries. Science and technology entertain some very tight and complex relations.

The sciences are fashioned by technologies. Conversely, today's technologies (IT, materials, life sciences and so on) are brimming over with science. Innovation is born from demand coming from the market (demand pull) as well as from imagination coming from research (technology push). Major innovations have sprung up independently of any form of science and many are the fruit of engineers and craftpeople referring to their usual technical universe. They rely on method-related elements and the way researchers go about their business rather than on their scientific statements as such.

Today, technologies use the sciences as one resource among others. Industrialists take on researchers in their R&D laboratories so that the company can assimilate published scientific information. Public authorities support transfer and popularisation centres to allow industrialists to understand the application possibilities of some fundamental research. They encourage the movement of ideas and the transfer of knowledge, part of which is tacit. Researchers create enterprises (start-ups) while at the same time pushing ahead with their scientific work. Looking at the history of innovations, it is impossible to draw any single conclusion as to the relations between discovery X and technological development Y. Two models qualify the relations between science and technology:

- **The hierarchical model**: science creates and proposes; technology appropriates and disposes. Technology uses science and is conditioned by it while science refers to nature and speaks on its behalf. Science comes first with its creations and discoveries, while technology comes second with its deductions and applications: *Nature → Science → Technology*. Science in itself contains potential technological applications.

- **The interactive model**: science and technology both invent and produce their own knowledge. Transfers are made via persons and in both directions while taking many detours.

The relations between technology and science are complex. An analysis of the relations between scientific publications, patents and the marketing of new products confirms this complexity together with the variability of science–technology interactions (Callon et al., 1986 (Box 8.5)).

Box 8.5   *The transistor*

Discovered in 1948 and applied in 1951: the relation seems obvious, direct and linear. And yet, the discovery in 1948 was preceded not only by many scientific events, but also by technological developments. It stemmed from work on quantum physics in 1932, which did not foresee the transistor effect. It also came from empirical work on semi-conductors that had been ongoing since their discovery in 1875, with the components being developed without the phenomenon being understood. It was also inspired by the fine-tuning of radar during the Second World War and new crystallisation and doping techniques in metallurgy. After the war, the researchers involved in this project found themselves working in the Bell laboratories. They used the conceptual tools of quantum physics and the recent technological breakthroughs in metallurgy. From this combination of scientific and technological work, the discovery of the transistor effect emerged in 1948. The application of this effect in 1951 was the result of some still very dissatisfactory tinkering. It was only after years spent developing complementary technologies that transistors began to produce reliable and controlled effects. In the case of the transistor, therefore, science was just one ingredient among many others.

Technology and Society

Technology and society are sometimes thought of as two spheres, one of which has an influence on the other. Thinking on this matter is dominated by four approaches: technical determinism, co-evolution of technology and society, social constructivism and the seamless web model.

- **Technical determinism** supposes that technological evolution is independent of society. It is autonomous, either because it is driven by an internal necessity, or because it is deduced from scientific development. It causes a social change; it is a force that is external to society and that weighs down on society. *Its impact* on society is therefore queried (see Ellul (1980) and the Technology Assessment movement). Innovation spreads thanks to its intrinsic properties. If technology is good, efficient, cost-effective and robust, it imposes itself on users who can do nothing else but adopt it. The question is how can society adapt to technological change?

- **Co-evolution of technology and society**. Simondon (1959) refers to coupling and co-evolution of the machine and its associated environment. This coupling stems from schemes used by the inventor to simultaneously apprehend both the object and its environment. Gille (1986) talks of a dual technological and social system as both technology and society are in a relationship of interdependence and compatibility. Mumford (1934) reports on a global co-evolution of technology and society where technology extends and strengthens organisational and political development. The pioneers of the French sociology of work, Georges Friedmann, Pierre Naville and Alain Touraine, studied the way in which technologies weigh down on the world of work and denounced the fatalistic attitude of bosses faced with the 'progress' of technology. They focused on the amount of initiative enjoyed by bosses when it comes to implementing technologies (idea of the social control of technologies):

  > Men had become mechanical before they perfected complicated machines to express their new bent and interest; and the will to order had appeared . . . in the monastery and the army and the counting-house before it finally manifested itself in the factory. (Mumford, 1934, p. 3)

- In **social constructivism**, technology is a materialised social relation. For Marx, the machine materialises a social rapport, which then imposes itself on workers (also see Noble, 1984 on digitally controlled machine tools). Feminist studies of technology analyse the representations of men and women and the more or less discriminatory strategies of those who fashion technologies (Cockburn and Ormrod, 1993; Cockburn and First-Dilic, 1994). They examine both the way in which technologies have an impact on relations between men and women and the way in which gender relations fashion technology.

- The **seamless web model**. For Hughes (1983), technological systems, such as electrification in the United States, are the result of many small inventions based on existing technologies in relation to circumstances. There is neither autonomy nor internal logic in the development of technology. However, its integration into systems imposes restrictions (given the stakes underlying economic and military competition), which guide the definition of problems and the technical solutions. Solving a technical problem is the same as solving an economic (or military) problem. There is no reason to distinguish between technology and society: there is only a seamless web. The actor-network theory proposes a similar analysis by reporting on the processes according to which sociotechnical networks are redefined.

Social Systems of Innovation

The relations between science, technology and society differ according to the social systems of innovation (see Chapter 1): scientific and industrial policies, R&D organisation and the status of researchers in companies, and the role of regulating authorities. The link between financing and politics is a central issue (Krige and Pestre, 1997). The amount of state intervention, which is stable in terms of volume, has become proportionally weaker with the economy's increasing dependence on science and technology. The situation has become paradoxical considering that scientists have become less well paid and have lost much of their prestige.

The links between teaching and research deserve special attention. University training systems are fashioned by a social demand that affects research logic. The number of lecturer/researcher positions and the inflow of high-level students in laboratories depend on the relative success of university courses. Research thus depends on the development of universities and on the social demands that mostly concern teaching.

Innovations are also linked to societal contexts that have an influence on dynamics: categories of actors (engineers, researchers and so on), forms of division of labour and the type of space where individuals acquire their qualifications. The strategies and dynamics of corporate innovation reflect these societal characteristics. In France, for example, innovation dynamics are partly random owing to the substantial difference between types of professional action logic (between salespeople, researchers and production engineers notably). To make innovative cooperation successful, industrialists have to invest in forms of organisation, such as project-based organisation (bringing together professionals from different functions). In Japan, the transfer between research and industrialisation is much easier because there is cooperation between hierarchical managers, who switch between different functions over the course of their career in a company.

Links between Science and Society

Several models (which are not exclusive with respect to each other) have been put forward to qualify the relations between science and society:

- The **confinement model** postulates an institutional separation between research (autonomous scientific community) and the rest of society.

- The **finalisation model** assumes that research is directed according to the objectives and priorities defined by society.

- The **entrepreneurial science model** is based on the idea that scientists are entrepreneurs of science. They develop and implement strategies that lead to the production of knowledge and to the capitalisation of this knowledge in economic and social terms (Etzkowitz, 2004 (in Chapter 2)).

> - The **triple helix or triangular model** is based on the hypothesis that science, industry and the State are closely and increasingly interlinked (Etzkowitz and Leydesdorff, 2000; Sábato, 1975 (in Chapter 2)).
>
> - The **distributed research model** suggests that the production of new knowledge stems from a heterogeneous set of actors for whom research is not the only activity.
>
> - The **knowledge regime model** reports on the plurality of ways in which scientific and socioeconomic actors are linked (Pestre, 2003, Shinn, 2002a).

Mode 1/Mode 2

According to Gibbons et al. (1994), the knowledge production mode has changed over time, moving from Mode 1 (confinement model) to Mode 2 (multiplication and distribution of knowledge production sites). Science is no longer the work of academic research centres alone (laboratories cut off from the rest of the world), but is spread across society: industry, consultancies, contract-based research companies, hospitals and governmental agencies thus also constitute places where knowledge is produced.

Furthermore, the proportion of the population trained in the sciences has increased. Most citizens receive some scientific training. Wherever they are employed, they channel this knowledge and the working methods emanating from the sciences and produce and formalise knowledge drawn from experience. Knowledge is subject to an increasing amount of attention from non-scientific actors.

The nature of knowledge is also taking on new forms: it is less a question of discovering fundamental principles that can be universally applied than of producing locally relevant knowledge. Cognitive and methodological resources as well as the criteria used to assess results have moved away from the academic traditions of science. The ability of a research result to be used by practitioners has become an important assessment criterion. The knowledge production mode has therefore moved from a science based on disciplines, governed by a hierarchy and isolated from society, to a science that is linked to society, is interdisciplinary and has new forms of organisation. Scientific actors are engaged in continuous negotiations to establish the relevance and legitimacy of their activity and obtain financing by applying to various organisations. This leads them to open up their field of interest and approach. The disciplinary divisions and distinctions between basic and applied research no longer hold as much weight.

These changes affect research and training institutions. Working with a diverse range of partners, such institutions now have to manage multiple functionalities. Teaching has to cater for a more diverse range of needs than in the past: intellectual training for a large share of the population, production of new knowledge, innovation, but also professional coaching of students and preparation for their entry into the working world. This means that old specialisations have to redefine themselves while everybody's role is becoming increasingly complex. In universities, professors have to lecture, do research and handle some of their institution's administrative work. On top of this, they also have to act as experts, follow their students during their placement periods in companies, and invite industrialists to come and lecture. Similar changes can be witnessed in large research organisations. In the United States, the National Science Foundation (NSF), for example, has redefined its end purposes by integrating social and economic issues. Managing scientific research independently of the problems facing society no longer seems acceptable.

This idea of a switch from one production mode to the other has been much debated in literature. Shinn (2002b), for example, shows that research has always operated according to several modes of financing and collaboration. Furthermore, the Mode 1/Mode 2 theory suggests the crumbling of borders between scientific institutions and the rest of the world, which is not confirmed by the facts. Peer validation via publications is still preponderant while most scientific knowledge stems from scientific logic that is intrinsic to specialities. The knowledge produced is not widely known and shared by society either. Nor is scientific content increasingly subject to partners and financial backers, notably industrialists (Krige and Pestre, 1997). The characterisation of current changes is still subject to discussion and requires much more research work.

The Scientist and the Non-Specialist

Another avenue to be explored in the social study of the sciences concerns the relations of citizens and students with respect to the sciences. Some authors have referred to students' loss of affection for scientific studies as well as the general public's change in attitude towards the sciences. According to this theory, society is keener to see that the sciences, expertise and decision-making processes do not escape democratic control, especially with respect to issues about the future and society. Various works, which will not be discussed in detail here, have also devoted their attention to the sociology of scientific training, to popularisation, scientific culture, scientific museology, social representations and attitudes of the public in relation to science and technology, and to the way science is treated in the media. We shall deal only with the question of scientific expertise.

The Expertise in Question

Expertise can be defined as a form of mediation requiring specialised knowledge and technology in order to meet a need to control innovation and its possible unexpected consequences (Granjou, 2004). This expertise has become problematic. The old 'rational control ideal', which had led to complex technological developments (aeronautical and nuclear), has been called into question. The disillusion comes from environmental problems (climate change, risks and so on) and is taking us towards ethics based on precaution and responsibility. It also stems from the failing of expertise, as in the contaminated blood affair, in which the legitimacy of an entire state was called into question, or mad cow disease, where the experts proved to be incapable of anticipating the consequences of innovation and ignored the danger signals. Scientific knowledge is questioned and called on to recognise its own relative indetermination. Climate modelling has shed doubt on the very possibility of making forecasts (Wynne, 1992). Furthermore, when scientific knowledge is transposed into society, its complexity has to be better taken into account. It is also important to be aware of the tacit norms and values underlying scientific content and approaches and the definition of problems.

The scandals and controversies have led to the setting up of new regulation systems, seen as 'counter models of expertise' compared with conventional scientific and technical expertise. Their role is to take into account the diversity of knowledge, including that of citizens (Wynne, 1992), and exercise control over the possible tacit content of specialised knowledge. Citizens' conferences show that non-specialists are able to grasp problems and difficult issues, without falling back on simplistic 'for' or 'against' positions. This leads to the question of what forms of citizen participation to include in the management of scientific and technological affairs. One of the main topics is the transparency of assessment, expertise and decision-making procedures. Simply stating that something is of a scientific nature has less and less impact on citizens, even when this statement comes from recognised experts. There is a growing obligation to explain the approaches used, so that they can be put to public opinion (Jasanoff, 2002).

The Problematics of Expertise in the Social Sciences

In sociological literature, the question of expertise was explored within the critical science movement, before becoming an autonomous issue (Granjou, 2003) relating to decision-making methods, the sociology of the sciences and the sociology of collective action.

The sociological criticism of expertise was first formalised in the work of Habermas (1970) who made a distinction between three decision-making models: the decisionist model (choice based on purely political criteria), the technocratic model (choice based on the objective knowledge of the constraints and facts known to the experts), and the pragmatic model (choice based on the dialogue between experts and non-specialists). The pragmatic model is the opposite of the technocratic model in which experts are brought in to back decisions, if not to hide the real motives, stakes and consequences. Analyses denounce the social role of experts in the engagement of nuclear programmes and in the hierarchical decisions to computerise companies. Recourse to contradictory expertise, to counter-expertise and to battles between experts opens an intermediary regulation space between science and the state. This affects the image of science and its claim to tell the truth, while at the same time providing scientists with new resources (Nelkin and Pollak, 1981).

The legitimacy of experts and their statements stems from their control of a body of specialised knowledge and the fact that they have been appointed by an institution. In France, experts act in the name of science and in the name of an institution and are therefore not required to explain the theories they put forward as part of their expertise. Hence, expertise stands apart from the dynamics surrounding the production of theoretical statements within scientific communities and within the Anglo-Saxon world where statements remain provisional, are subject to discussion and put to the test. When there is uncertainty, 'expertise should open the discussion not close it'.

There is also the question of the legitimacy and robustness of knowledge and expertise, as well as that of the efficacy of using such knowledge. Roqueplo (1997) argues for clear borders to be established between the expert and the decision maker so that the expertise provided is independent and there is no confusion of roles. Expertise is the fruit of a contradictory discussion between specialists. This discussion is nourished and supported by an ethic of the objectivation which enables the conclusions to be validated. Expertise is only valid for a provisional and limited period while decisions exceed the boundaries of knowledge. In accordance with the precaution principle (dissociation between expertise and decision), expertise cannot automatically bend a decision. On the contrary, for Callon and Rip (1992), the fact that experts' statements are the result of ties drawn between different types of consideration promotes the argument for an increased number of viewpoints and a widening of the social debate. It is a question of opening up the expertise process to groups with specific local knowledge and interests. Such hybrid forums produce conclusions that are even more robust because they are the result of democratic confrontation; objectivation alone is not enough to ensure robustness. The hybrid forum is opposed to the dual delegation of power reflected in representative democracy and the 'Great Divide' between experts and non-specialists (Callon et al., 2001). The specificity of experts is thus tending to disappear. From a practical point of view, the last few decades have witnessed the multiplication of new deliberative practices concerning scientific and technological stakes and practices. These practices break away from the French tradition whereby decision making is monopolised by elected officials with the help of technical experts. They rely instead on the mobilisation of scientific and non-scientific actors, promoters of interests, stakes and knowledge of all kinds in order to outline problems and invent solutions (Latour, 2004, Marris et al., 2005, 2008, Joly, 2007, Bonneuil et al., 2008). The contribution of non-scientists reflects a breakaway from the public education model according to which the exacerbated perception of risks comes from a lack of scientific culture on behalf of the non-specialist public, with the solution being for experts to disseminate suitable information.

Although several authors underline the contribution of non-scientists, others wonder what room is left for scientific and technical expertise (Weingart, 1999) and focus on restoring the borders (Collins and Evans, 2002). They argue for the renewal of the distinction between expert and non-specialist with rights and specific responsibilities (Millstone and van Zwanenberg, 2000). They promote the need for a normative theory of expertise to overcome the dual pitfall of scientism and relativism. Jasanoff (1987, 1990) shows that actors unceasingly strive to construct and deconstruct the borders between science and politics. Granjou (2004) suggests analysing the forms of arbitration and distinction between types of knowledge and the proximity between experts and non-specialists. Granjou queries the relevance of the expert category by looking at the way in which these experts and expert committees conceive and manage their missions, responsibilities and asymmetries relating to knowledge and experience. She shows that experts do not fulfil a role that is predetermined by their mandate or their prior qualifications (skills, experience, reference and belonging), or by their representation of what is supposed to be good expertise. The role of the expert and the content of an expert's work is (re)constructed over the course of their action. Both are subject to learning, appropriation, negotiations between experts, experimentation and thinking before being transformed into cognitive frameworks common to the members of the committee. Experts put their mandate to the test, set up a form of collegiality and a sharing of tasks, discover the uncertainty linked to certain types of knowledge, query the notion of proof, and develop an awareness of information asymmetries. Experts quit this role when the work does not feed their own scientific interests. They negotiate role-sharing among themselves and with the decision makers. To do this they underline their scientific background. They protect themselves against the risk of representational legitimacy being usurped. They develop a variety of references to the uncertain, using specific theoretical modelling: they point out the limits of validity of results, they use negative formulations such as 'We cannot exclude the fact that . . .' they underline the total absence of data, and express certain divergences with respect to other experts. In fact, they displace the rational-legal model by taking into account the objections proffered by non-specialists. Their competency comes less from the corpus of knowledge they control than from their experience with the mechanisms of objection and objectivation at work in their speciality, with the stubbornness of reality and that of their colleagues:

> It would be difficult to find a substitute for the expert in a role where they are called on to relativise knowledge, qualify hypotheses and restore the conditions needed to validate interpretations as they review the handiwork carried out in laboratories, for the benefit of non-specialists, in order to build a theoretical model and a solid hypothesis within specific limits. (Granjou, 2004, p. 404)

In the institutionalist tradition there is an obvious difference between scientists and others. In the relativist tradition, this distinction had disappeared or was considered illusory. The neo-institutionalist movement suggests reintroducing this demarcation. For the actor-network theory, these borders are neither denied nor presupposed. The important thing in this theory is to analyse the concrete forms of hybridisation and dissociation, the mixtures and borders implemented by the actors and what this produces.

Science and Society: The Question of Democracy

Over the centuries, science and the definition of its place within society have been affected by antagonistic movements. Attitudes to science have varied from trust to mistrust, while the question of their democratic structuring is ever present.

Changing Attitudes towards the Sciences

No one period has seen a single prevailing attitude with respect to the sciences. In the seventeenth century, scientists' movements came up against the political and religious institutions in place. The Bacon programme, relating to the experimental sciences, was welcomed by part of British society as a 'real prophecy' that would enable a public consensus to be maintained. Science was supposed to lead to social peace, in spite of the political and theological implications of the issues dealt with, thanks to an agreement about the research procedures relative to scientific problems. Science was seen as a non-violent and civilised means of solving conflicts, that is, it did not involve power struggles. Although this conception of science and the government's ideal based on science were welcomed, there was nevertheless no consensus on the matter. Institutions and social groups did not view this change in the method of solving society's problems and managing social evolution in a good light. In the nineteenth century, alongside the now dominant movement based on positivist philosophy and confidence in scientific progress, there was an anti-science movement promoting individual sensitivity and romanticism. At the beginning of the twentieth century, the Roman Catholic Church denounced modernism. Thinkers like Joseph Renan responded by praising the role of science in society and underlining its undeniable benefits.

Throughout the twentieth century, the development of science and technology took place at the same time as a progressive dissociation between ideas relating to scientific progress and to human progress. After the First World War, German society accused Newtonian science of having led the nation into the world of industry and weapon-making and into a race for power that finished badly. It denounced scientific, technical and industrial progress and called for a return to romantic values. This crisis nevertheless only lasted a few years. Furthermore, until the Second World War, scattered voices could be heard as they tried to spread the message that scientific and industrial progress did not necessarily lead to moral and social progress. This criticism was welcomed by some when it concerned work in factories (Frédéric Le Play on the worker's condition, Marx on machinism, and Friedmann on piecemeal work following Taylorism). In spite of this sparse criticism, science was safe: the myth of the brilliant and disinterested scientist prevailed in society, taking on the face of Einstein; great scientific institutions were created. Merton showed that science was governed by norms of disinterestedness.

However, when the atomic bomb was dropped on Hiroshima and Nagasaki in 1945, this sounded the knell of the relationship of confidence between science and society. The image of science was no longer that of an autonomous sphere of activity fulfilling an obvious social role: pacification and improved reason and morals. It was no longer 'free of any social responsibility' (Pestre, 1984). Hiroshima made people think that a secret pact had been signed between scientists and the political and military powers behind their back. There then emerged, at the very heart of the physicists' community, a critical movement opposed to any form of connivance between science and the military: the Pugwash movement. The event left lasting scars on the collective consciousness and fired the STS (science, technology and society) movement in the 1970s. This movement explored the possibility of setting up a social control of technologies.

This questioning of the sciences was nevertheless limited. The context of the cold war brought scientists and the military closer: economic development conferred a certain amount of technical comfort on households in Western countries; in rural areas the productivist model caused a boom in agricultural production; the level of school education rose while many young people headed towards a career in science and technology.

It was not until a new crisis arose (that is, the oil crisis in 1973) that the science debate, which had been upheld by marginal groups in the meantime, was put back into the limelight. The sudden rise in oil prices increased awareness of the limits to the exploitation of nature by industrial society. Scientific and technological progress was once more pointed at with an accusing finger, not because of a handful of 'bad applications' but because it was seen as a victim of its own success. The accusations also extended to the hypernucleation of the planet, agricultural surplus in Western countries, chemical pollution, global warming, oil slicks, acid rain, pesticide residue in food, and the hole in the ozone layer. 'Stop Growth' and 'Zero Growth' became the new slogans. Besides questions focusing on the development model, the role and place of science in society was also targeted. Science was no longer considered as 'fundamentally good'. Going beyond specific problems, in fact, a new representation of science and technology emerged. After having been considered, up until then, as humanity's allies, helping human beings to survive the wild, natural world, science and technology came to be seen as a source of uncertainties (Nowotny et al., 2001), risks and concerns. Science was viewed as a suspicious and conniving activity; scientists were compromised and the trust was broken. However, surveys on citizens, whether on the subject of GMOs or nanotechnologies, showed that there was no real disavowal of research. Support for the French movement 'Let's save research' in 2003–04 confirms the credit bestowed by the population on the disinterested research model, designed to serve society.

From Criticism of Technoscience to Science's Democratic Calling

In the field of social and human sciences, a critical movement with respect to science emerged. Thinkers (including Martin Heidegger) denounced the subjecting of science to projects that were far from contemplative or liberating: perceived as an instrument to control objects via thought, aiming to understand and control the natural world, science was enrolled in a project to desacralise and dominate nature. Marcuse (1964) and Habermas (1970, 1971) denounced the domination of a single instrumental rationality, extending to all levels of society, annihilating other forms of thought and acting as a form of insidious political domination. According to Habermas, axiological neutrality (in terms of values) leads to practical efficacy, but also to the illusion of a pure theory that prevents us from being aware of the close links between scientific project and societal stakes. Founded on its neutrality and extraterritorial nature, scientific authority is thus an obstacle to critical thinking about science itself.

In the 1970s, researchers developed self-criticism of the sciences (Lévy-Leblond and Jaubert, 1972; Roqueplo, 1974; Rose and Rose, 1970). The main targets of this criticism were scientists' involvement in the arms race, the myth of science as the only means of supplying objective and non-mythical knowledge, the ideology implicit in the analytical approach, the fragmentation of knowledge and the inability of scientists to develop critical thinking about the directions of science. Researchers launched the idea of 'science shops' so that scientific knowledge could serve the problems of society and citizens. This did not slow down investments in research during this period. Public research programmes were launched in order to find solutions to the new problems facing society, for example, research into alternative energies that were not based on oil or nuclear power. At the turn of the 1980s, science and technological innovation were called on to bring nations out of the economic crisis that reigned.

In Great Britain, criticism of the sciences stemmed from the relativist sociology that targeted the hegemonic position of physics and revealed the power struggles and social interests underlying the production of knowledge. Feminist critics denounced the male chauvinism that had slipped into so-called 'objective knowledge contents'. Simultaneously, authors developed a critical analysis of techno-scientific development (Ellul, 1980). Furthermore, several accidents occurred (including Seveso and Three Mile Island), forging the notion of 'major technological risks' (Lagadec, 1982). This period also marked the start of the massive spread of computers in companies and administrative offices, and the setting up of a moratorium of molecular biologists with respect to the use of recombinant DNA (genetic engineering). In both the United States and Europe, institutions and methods for the societal evaluation of technologies and their impacts on society emerged. Their aim was technology assessment. Critical thinking was organised within public institutions, developing a corpus of knowledge that shed a different view on science and technology. This then raised questions about the political management of technologies, the assessment of their positive or harmful effects, the unequal sharing of these effects on populations (present and future), the illusion of 'objective constraints' and the partial nature of choices. Finally, the question of technical democracy and society's control over the political decisions behind science and technology emerged.

In the 1980s, criticism focused upon nuclear power, IT and biotechnology: denucleation and creation of ethics committees. At the same time, nations engaged in a race for innovation in order to overcome the economic crisis. Recounting the development of computer-assisted design (CAD) tools, the anthropologist Downey (1992) published an article reflecting the view that 'CAD/CAM saves the nation'. Sociologists of the sciences joined in the thinking about technical democracy, the societal assessment of technologies (*constructive technology assessment*), but also in the management of research organisations and innovation.

In the 1990s, a series of controversies broke out about technological R&D

activities: the contaminated blood affair, mad cow disease, global warming, GMOs and, as of 2003, nanotechnologies. Alter-globalist movements protested against capitalistic-based logic and the frantic race for innovation, while radical groups campaigned to 'stop everything'. Social scientists studied these socio-scientific controversies, attempting to understand their underlying forces and dynamics, and even trying to attract the attention of institutions with respect to the extent of the dynamics at work in civil society. The question of the risks and uncertainties linked to science and technology gained importance.

Science and expertise, which were supposed to ensure social peace through established truths, only complicated the debates instead of simplifying them. Thinking centralised around the question of expertise, the societal regulation of risks and the controversies relating to science and technology (Collins and Evans, 2002). Latour (2004) referred to the parliament of things and wondered how to get the sciences to answer the call of democracy. The important questions at the start of the third millennium concentrated on the conditions pertaining to the debate and to democratic control, the concrete forms of knowledge demonstration and mobilisation (experts and non-specialists) and the procedures for representation and participation. The theme of the 2004 Society for Social Studies of Science (4S) seminar was a perfect reflection of this concern: 'Public Proofs: Science, Technology and Democracy'.

In a society that is now known as the 'knowledge society', it is difficult to imagine that decisions concerning all of society and future generations should be taken by experts alone. This is all the more surprising considering that citizens have become increasingly dependent on scientific knowledge, which is itself incomplete, provisional and uncertain. The problem is all the more poignant considering the ever-changing nature of the risks: from the risk of an accident occurring to its identifiable consequences (major technological risks), the problem is that of the widespread consequences and health-related risks. New fears have appeared concerning the possible risks inherent in nanoparticles and nano-objects, such as a nanorobot able to reproduce itself like a virus. The notion of risk itself has been called into question because it assumes that it is possible to distinguish scientific and technological knowledge, on the one hand, and its involuntary societal effects, on the other. Can anybody say where the boundary actually lies?

Notes

1 In some cases, the text has already been presented many times, leading to amendments but also to a certain notoriety before publication.
2 Except for Zuckerman and Merton (1971) mainly, on the overall system of evaluation, and the analysis of Vilkas (1996) on the French National Centre for Scientific Research (CNRS) national committee.
3 The notion of 'entity' is used here because it is neutral. Latour (1987) uses the semiotic notion of '*actant*'. These two notions are more appropriate than that of 'actor', normally defined as being human.

Recommended Reading

References appearing in other chapters: Latour (1996) in Introduction; Eisenstein (1979) in Chapter 3; Callon (1986), Rabeharisoa and Callon (2002) in Chapter 6; Vinck (1992) in Chapter 7.

Callon, M., Lascoumes, P. and Barthe, Y. (2001), *Agir dans un monde incertain. Essai sur la démocratie technique*, Paris: Seuil.

Callon, M., Law, J. and Rip A. (1986), *Mapping the Dynamics of Science and Technology*, Basingstoke: Macmillan.

Latour, B. (1987), *Science in Action: How to Follow Scientists and Engineers through Society*, Cambridge, MA: Harvard University Press.

Latour, B. (1988), *The Pasteurization of France*, Cambridge, MA: Harvard University Press.

Latour, B. (1993), *We Have Never Been Modern*, Cambridge, MA: Harvard University Press.

Latour, B. (2004), *Politics of Nature: How to Bring the Sciences into Democracy*, Cambridge, MA: Harvard University Press.

Latour, B. (2005), *Reassembling the Social: An Introduction to Actor-Network Theory*, Oxford: Oxford University Press.

References

References appearing in other chapters: Ashmore et al. (1989) in Introduction; Sábato (1975), Zuckerman and Merton (1971) in Chapter 2; Gibbons et al. (1994), Krige and Pestre (1997) in Chapter 3; Nowotny et al. (2001), Vilkas (1996) in Chapter 4; Callon et al. (1986) in Chapter 5; Knorr-Cetina (1999) in Chapter 6; Knorr-Cetina (1981), Pontille (2003), Star and Griesemer (1989) in Chapter 7.

Amin, A. and Cohendet, P. (2004), *Architectures of Knowledge: Firms, Capabilities and Communities*, Oxford: Oxford University Press.

Bonneuil, C., Joly, P.B. and Marris, C. (2008), 'Disentrenching experiment? The construction of GM-crop field trials as a social problem in France', *Science Technology and Human Values*, **33** (2), 201–29.

Callon, M. (1986), 'The sociology of an actor-network: the case of the electric vehicle', in M. Callon, J. Law and A. Rip (eds), *Mapping the Dynamics of Science and Technology*, Basingstoke: Macmillan, pp. 19–34.

Callon, M. and Law, J. (1982), 'On interests and their transformation: enrolment and counter-enrolment', *Social Studies of Science*, **12** (4), 615–25.

Callon, M. and Rip, A. (1992), 'Humains, non-humains. Morale d'une coexistence', in J. Theys and B. Kalaora (eds), *La Terre outragée. Les experts sont formels*, Paris: Autrement, pp. 17–31.

Cockburn, C. and First-Dilic, R. (1994), *Bringing Technology Home: Women, Gender and Technology*, Milton Keynes: Open University Press.

Cockburn, C. and Ormrod, S. (1993), *Gender and Technology in the Making*, London, Sage.

Collins, H. and Evans, R. (2002), The third wave of science studies: studies of expertise and experience', *Social Studies of Science*, **32** (2), 235–96.

Deloncle, G. (2004), 'Les logiques sociales de la critique scientifique en sociologie. Une analyse des compte-rendu négatifs dans la *Revue française de sociologie*, Congrès AISLF, Tours, juillet 2004. Available at: http:docs. google.com/gview?a=v&q=cache:cTg71WyLYN0J:mshe.univfcomte.fr/programmation/col04/documents/preactes/Deloncle.pdf+Geoffrey+Deloncle&hl=fr&gl=fr

Desrosières, A. (2002), *The Politics of Large Numbers: A History of Statistical Reasoning*, Cambridge, MA: Harvard University Press.

Downey, G. (1992), 'CAD/CAM saves the nation? Toward an anthropology of technology', *Knowledge and Society*, **9**, 143–68.

Ellul, J. (1980), *The Technological System*, New York: Continuum.

Epstein, S. (1995), 'The construction of lay expertise: AIDS activism and the forging of credibility in the reform of clinical trials', *Science, Technology and Human Values*, **20**, 403–37.

Etzkowitz, H.I. and Leydesdorff, L. (2000), 'The dynamics of innovation: from national systems and "mode 2" to a triple helix of university–industry–government relations', *Research Policy*, 29 (2), 109–23.

Gille, B. (1986), *The History of Techniques*, New York: Gordon and Breach Science Publishers.

Granjou, C. (2003), 'L'expertise scientifique à destination politique', *Cahiers Internationaux de Sociologie*, **114**, 175–83.

Granjou, C. (2004), 'La gestion du risque: entre technique et politique. Comités d'experts et dispositifs de traçabilité à travers les exemples de la vache folle et des OGM', PhD in Sociology, Université René Descartes (Paris V), Paris.

Habermas, J. (1970), 'Technology and science as ideology', in J. Habermas, *Toward a Rational Society: Student Protest, Science, and Politics*, Boston, MA: Beacon Press, pp. 81–122.

Habermas, J. (1971), *Knowledge and Human Interests*, Boston, MA: Beacon Press.

Hughes, T. (1983), *Networks of Power: Electrification in Western Society, 1880–1930*, Baltimore, MD and London: Johns Hopkins University Press.

Jasanoff, S. (1987), 'Contested boundaries in policy relevant science', *Social Studies of Science*, **17**, 195–230.

Jasanoff, S. (1990), *The Fifth Branch: Science Advisers as Policymakers*, Cambridge, MA: Harvard University Press.

Jasanoff, S. (2002), 'Citizens at risk: cultures of modernity in the US and EU', *Science as Culture*, **11** (3), 363–80.

Joly, P.B. (2007), 'Scientific expertise in public arenas – lessons from the French experience', *Journal of Risk Research*, **10** (7), 905–24.

Lagadec, P. (1982), *Major Technological Risk: An Assessment of Industrial Disasters*, London: Elsevier.

Latour, B. (1982), 'Le centre et la périphérie à propos des transferts de technologies', *Prospective et santé publique*, **24**, 37–44.

Latour, B. (1983), 'Give me a laboratory and I will raise the world', in K. Knorr-Cetina and M. Mulkay (eds), *Science Observed*, London: Sage, pp. 141–70.

Law, J. (1987), 'Technology, closure and heterogeneous engineering: the case of the Portuguese expansion', in W. Bijker, T. Pinch and T.P. Hughes (eds), *The Social Construction of Technological Systems*, Cambridge, MA: MIT Press, pp. 111–34.

Lévy-Leblond, J.M. and Jaubert, A. (1972), *(Auto)critique de la science*, Paris: Seuil.

Marcuse, H. (1964), *One-dimensional Man: Studies in the Ideology of Advanced Industrial Society*, Boston, MA: Beacon.

Marris, C., Joly, P.B. and Rip, A. (2008), 'Interactive technology assessment in the real world: dual dynamics in an iTA exercise on genetically modified vines', *Science, Technology and Human Values,* **33** (1), 77–100.

Marris, C., Joly, P.B., Ronda, S. and Bonneuil, C. (2005), 'How the French GM controversy led to the reciprocal emancipation of scientific expertise and policy making', *Science and Public Policy*, **32** (4), 301–8.

Millstone, E. and van Zwanenberg, P. (2000), 'Beyond skeptical relativism: evaluating the social constructions of expert risk assessments', *Science, Technology and Human Values*, **25** (3), 259–82.

Mumford, L. (1934), *Technics and Civilization*, New York: Harcourt Brace.

Nelkin, D. and Pollak, M. (1981), *The Atom Besieged: Extraparliamentary Dissent in France and Germany*, Cambridge, MA: MIT Press.

Noble, D. (1984), *Forces of Production: A Social History of Industrial Automation*, New York: Knopf.

Oudshoorn, N. (1994), *Beyond the Natural Body: An Archaeology of Sex Hormones*, London: Routledge & Kegan Paul.

Pestre, D. (1984), *Physique et physiciens en France (1918–1940)*, Paris: Éditions des Archives Contemporaines.

Pestre, D. (2003), 'Regimes of knowledge production in society: towards a more political and social reading', *Minerva*, **41** (3), 245–61.

Roqueplo, P. (1974), *Le Partage du savoir: science, culture, vulgarisation*, Paris: Seuil.

Roqueplo, P. (1997), *Entre savoir et décision. L'expertise scientifique*, Paris: INRA.

Rose, H. and Rose, S. (1970), *Science and Society*, London: Penguin.

Shinn, T. (2002a), 'The transverse science and technology culture: dynamics and roles of research technology', *Social Science Information*, **41** (2), 207–51.

Shinn, T. (2002b), 'The triple helix and new production of knowledge: prepackaged thinking on science and technology', *Social Studies of Science*, **32** (4), 599–614.

Simondon, G. (1959), *On the Mode of Existence of Technical Objects*, London, ON: University of Western Ontario Press.

Weingart, P. (1999), 'Scientific expertise and political acountability: paradoxes of sciences in politics', *Science and Public Policy*, **26** (3), 151–61.

Wenger, E. (1998), *Communities of Practice, Learning, Meaning and Identity*, Cambridge: Cambridge University Press.

Wynne, B. (1992), 'Science and social responsibility', in J. Ansel and F. Wharton (eds), *Risk: Analysis, Assessment and Management*, New York: John Wiley, pp. 137–52.

Conclusion

This manual has allowed us to explore a wide variety of approaches and issues relating to the study of science. It has underlined the importance and the thorny nature of the ever-increasing questions it raises. Some of these stem from the very dynamics of the sciences, as practices change according to the objects, instruments and forms of organisation involved. Moreover, they are inherently tied to what is happening in society today and the challenges facing it. Nevertheless, there are other questions arising from changes to the social study of science itself. The concluding paragraphs review some of the structural elements relating to these questions.

Questions Arising from Recurrent Academic Debates

In the debate about the relative autonomy of the institution of science and the independence of scientific knowledge with respect to social influences, some authors reaffirm the idea of a partially autonomous, immanent development of science. They underline the differential role of social and cognitive factors (or epistemic factors), and reject or reformulate the postulates of relativist and constructivist sociology. They develop a neoinstitutionalist sociology (Kreimer, 1997) or a socio-epistemology. Boudon and Clavelin (1994), for example, state that the position of Karl Mannheim, according to which certain (scientific) proposals are independent of the social context, is the only reasonable position to adopt. It recognises the influence of social factors on scientific development, but defends the idea that science is intrinsically objective. From this standpoint, the actor-network theory (ANT) is reproached for its inability to differentiate and counterbalance the influence of different factors (social, cognitive and so on). However, given that ANT does not accept the idea of a predefined space of causes, this accusation holds no weight.

In the debate about the establishment of norms to be used to judge scientific practices and manage researchers, ethnomethodologists are pitched against the partisans of social epistemology, the aim of which is to find ways to say what should be done in science. To do this, according to Fuller (1988), a certain detachment is required when studying scientific actors and their communication schemes, which need to be analysed in an even more scientific way using psychology (Shadish and Fuller, 1994).

In the debate about the role of the social context, relativists and ethnomethodologists confront each other about how to interpret Wittgenstein and his notion of rules. For Bloor, the fact that a rule does not state how it should be applied justifies the use of a social type of causality to bridge the gap. For Lynch, there is no room for reductive social concepts since things are linked up in effective practices.

In the debate about the role of nature, relativists are opposed to supporters of the actor-network theory and ethnomethodology, who are accused of naturalistic regression (reintroduction of a natural causality).

In the debate about the critical role of the sociology of science, relativists recommend that this sociology be involved in the public debate denouncing the hegemony of the natural sciences. Their reproach lies with the amusing, if not epistemologically radical, written tricks of reflexivists, which they say are politically impotent and devoid of any message. The reflexivist Woolgar, on the other hand, considers that by aiming to criticise the imposture of the natural sciences, relativist sociology adopts this same imposture. The same debate focuses on the theoretical and methodological relevance of the generalised symmetry principle. Although the position is epistemologically radical, it is also politically regressive. The critical role of the sociology of scientific knowledge is lost in it.

In the debate about discriminations and power within the sciences, feminist sociologists reproach constructivist approaches for their inability to report on the discriminations operated within and by science and technology. The fact that there is no 'gender' analysis category prevents these researchers from measuring the social handicaps of female scientists and the embedding of social gender relations in scientific and technological content. Other voices can be heard in this debate denouncing the lack of consideration of domination-related questions.

The Question of Reflexivity and the Anthropological Foundation of the Sciences

Many sociology of science schools nevertheless have a point in common: they think they provide an adequate description and representation of the sciences. However, Woolgar's reflexivity queries whether this is not simply an illusion.

The things of the world (nature, society and so on) and their scientific representations appear to belong to two different worlds; the first are independent of the second. For reflexivists, this construction and constitution of the world in two parts is the result of the work of scientific writing. Scientists establish a moral order (a distribution of beings with different statuses) and an ideology of representation (scientists simply talk on behalf of and are dictated to by nature).

The sociology of the sciences does the same with respect to the sciences, that is, it represents them. As it does so, it shares the same ideology of representation and the same moral order as that of the scientists it studies. Just like natural scientists, sociologists would have us believe that they analyse and describe a reality that they themselves have not built, but which they simply observe and explain. They fall into the same representative ideology as the scientists themselves, which means that they cannot reveal its hidden depths. Even the most naive observer is unable to produce the necessary distance. Instrumental ethnography produces stories in an attempt to unearth things unknown to the reader. It demystifies scientific work. In fact, ethnographic reports imply a sort of sociological irony; their form (a serious academic presentation) contrasts with what happens in the laboratory (disorder, tinkering and negotiations). They fail to question the core of scientific activity, that is, the notion of representation. They put forward two solutions: (i) the transcription of conversations between scientists, processed using conversation analysis methods; and (ii) the description of the way in which scientists construct their reports.

The *reflexive* ethnography promoted by Woolgar, on the other hand, has a strategic role in that it offers the opportunity to reflect on and better understand certain aspects of our culture. It is about exploring our own use of representation by exploring various forms of literary expression where the problem of representation constantly captivates the reader. It is about making readers aware of their own involvement in the text by underlining its fictional nature. Mulkay thus proposes a fiction based on the Nobel Prize award in which dissonant voices can be heard, that is, those of characters who normally stay quiet (the spokesperson of those who have not received the prize and the prize winner's spouse). It is a question of uncovering representation tactics and devices. The authority of scientific representation comes from the fact that the authors of such representation silence others. Once this exclusion has been uncovered, the text comes across as an artificial construction and not as discourse reflecting the facts. One solution consists in varying the voices and literary forms, as Ashmore et al. (1989) do for health economics or Latour (1996) about the Aramis underground. Traweek (1988) prefers to speak in a single voice, but she stages herself as an author in her essay on high energy physics in Japan and the United States. Latour (1988) puts forward the idea of 'infra-reflexivity': instead of writing about how to write and posting methodological warnings, the idea is simply to write, present a point of view (the author is part of the network and story they are studying) in a style that makes it reflexive. What he actually suggests is 'methodological deflation'.

To probe the cultural depths of scientific practice, sociologists must thus remove themselves from their scientific perspective and, for example, use the scientific practice observed to query their own practice as observers. It is no longer a question of using reliable and neutral techniques in order to show reality (that of the laboratory, for example) as it is, but of being in the laboratory and inviting the reader to question the practice of carrying out surveys, of being an observer and of being observed. From this point of view, science is not reflexive. It is a language that hides and denies its linguistic nature. It is a social practice and a construction in self-denial. Reflexive sociology aims to break this illusion.

Similarly, anthropology queries the foundation of scientific activity and its demarcation with respect to other social activities. Scientific work is built upon a distinction between scientific and non-scientific facts (for example, events and choices in the private life of researchers). Understanding science thus requires an understanding of this anthropological foundation, notably that of the demarcation (Hernandez, 2001) and of the parenthesis (Thill, 1973) that constitutes the laboratory. It also involves understanding how people use this demarcation differently, depending on whether they are men or women, scientists or technicians.

Recommended Reading

References appearing in other chapters: Hernandez (2001), Thill (1973), Traweek (1988) in Chapter 7.

Latour B. (1988), 'The politics of explanation: an alternative', in S. Woolgar (ed.), *Knowledge and Reflexivity, New Frontiers in the Sociology of Knowledge*, London: Sage, pp. 155–76.

References

References appearing in other chapters: Ashmore et al. (1989), Latour (1996) in Introduction.

Boudon, R. and Clavelin, M. (eds) (1994), *Le Relativisme est-il résistible? Regards sur la sociologie des sciences*, Paris: Presses Universitaires de France.
Fuller, S. (1988), *Social Epistemology*, Bloomington, IN: Indiana University Press.
Kreimer, P. (1997), 'L'universel et le contexte dans la recherche scientifique', Doctoral thesis, CNAM, Paris.
Shadish, W. and Fuller, S. (1994), *The Social Psychology of Science*, New York and London: Guilford Press.

Appendix

Relevant Journals

Bulletin of Science, Technology and Society
Public Understanding of Science
Radical Science Journal, Science for People
REDES (*Revista de estudios sociales de la ciencia*)
Research Policy
Revue d'Anthropologie des Connaissances
Science as Culture
Science in Context
Science Studies
Science, Technology and Human Values
Science Technology and Society
Social Studies of Sciences
Technology and Culture
Technology in Society
Technoscience

Scientific Associations

4S – Society for Social Studies of Science: publisher of *Science, Technology and Human Values* (STHV) and *Technoscience*, http://www.4sonline.org/.

AFS – Thematic network 29 'Sociologie des sciences, des techniques et de l'innovation' of the French Sociological Association.

AISLF – Research Committee 29 'Sciences, innovation technologique et société': International Association of French Speaking Sociologists: it is involved in the publication of the *Revue d'Anthropologie des Connaissances*, http://www.univ-tlse2.fr/aislf/gt6/index1280.htm.

EASST – European Association for the Study of Science and Technology: it publishes *Social Studies of Sciences*, http://www.easst.net/.

ESOCITE – Estudios sociales de la ciencia y tecnologia: Latin American association.

ESST – Inter-university European Association on Society, Science and Technology: it is responsible for a European Masters, http://www.esst.uio.no/.

ISA–RC 23: Research Committee on Sociology of Science and Technology of the International Sociological Association, http://www.ucm.es/info/isa/rc23.htm.

SAC – Société d'Anthropologie des Connaissances.

SHOT – Society for History of Technology: it publishes *Technology and Culture*, http://shot.press.jhu.edu/.

SSTNET – Sociology of Science and Technology Network: European platform related to the European Sociological Association (ESA), http://sstnet.iscte.pt/.

STS.CH – Swiss Association for Social Studies of Sciences, http://www.sts.unige.ch/.

Acronyms

AERES:	Agence d'Evaluation de la Recherche et de l'Enseignement Supérieur
AII:	Agence de l'Innovation Industrielle
ANR:	Agence Nationale de la Recherche
CEA:	Commissariat pour l'Energie Atomique
CERN:	Organisation européenne pour la recherché nucléaire (initially "Conseil Européen pour la Recherche Nucléaire")
CIFRE:	Conventions Industrielles de Formation par la REcherche
CNAM:	Conservatoire National des Arts et Métiers
CNRS:	Centre National de la Recherche Scientifique
CNU:	Conseil National des Universités
CREST:	Comité de la Recherche Scientifique et Technique
EDF:	Electricité De France
INRA:	Institut National de Recherche Agronomique
INRETS:	Institut National de Recherche sur les Transports et leur Sécurité
INRIA:	Institut National de Recherche en Informatique et Automatique
INSERM:	Institut National de la Santé et de la Recherche Médicale
IRAM:	Institut de Radioastronomie Millimétrique
IRD:	Institut de Recherche pour le Développement
IRIA:	Institut de Recherche en Informatique et Automatique
LETI:	Laboratoire d'Electronique et de Technologies de l'Information
ONRS:	Office National des Recherches Scientifiques et industrielles et des inventions

Names index

Subject index